ALLE·ZEIT·WACH·1842

Communications and Control Engineering Series
Editors: A. Fettweis · J. L. Massey · M. Thoma

J. A. Richards

Analysis of Periodically Time-Varying Systems

With 73 Figures

Springer-Verlag Berlin Heidelberg New York 1983

J. A. RICHARDS
School of Electrical Engineering
and Computer Science
University of New South Wales
P.O. Box 1
Kensington, N.S.W. 2033, Australia

ISBN-13:978-3-642-81875-2 e-ISBN-13:978-3-642-81873-8
DOI: 10.1007/978-3-642-81873-8

Library of Congress Cataloging in Publication Data
Richards, John Alan, 1945–.
Analysis of periodically time-varying systems.
(Communications and control engineering series)
Bibliography: p. Includes index. 1. System analysis.
I. Title. II. Series.
QA402.R47 1983 003 82-5978 AACR2

Typesetting: Asco Trade Typesetting Ltd., Hong Kong

2061/3020-543210

To Dick Huey

Preface

Many of the practical techniques developed for treating systems described by periodic differential equations have arisen in different fields of application; consequently some procedures have not always been known to workers in areas that might benefit substantially from them. Furthermore, recent analytical methods are computationally based so that it now seems an opportune time for an applications-oriented book to be made available that, in a sense, bridges the fields in which equations with periodic coefficients arise and which draws together analytical methods that are implemented readily. This book seeks to fill that role, from a user's and not a theoretician's view.

The complexities of periodic systems often demand a computational approach. Matrix treatments therefore are emphasized here although algebraic methods have been included where they are useful in their own right or where they establish properties that can be exploited by the matrix approach. The matrix development given calls upon the nomenclature and treatment of H. D'Angelo, Linear Time-Varying Systems: Analysis and Synthesis (Boston: Allyn and Bacon 1970) which deals with time-varying systems in general. It is recommended for its modernity and comprehensive approach to systems analysis by matrix methods. Since the present work is applications-oriented no attempt has been made to be complete theoretically by way of presenting all proofs, existence theorems and so on. These can be found in D'Angelo and classic and well-developed treatises such as

McLachlan, N. W.: Theory and application of Mathieu functions. Clarendon: Oxford U.P. 1947. Reprinted by Dover, New York 1964.
Arscott, F. M.: Periodic differential equations. Oxford: Pergamon 1964.
Magnus, W.; Winkler, S.: Hill's equation. New York: Wiley 1966.

Instead, this book relates theory to applications via analytical methods that are, in the main, computationally-based.

The book is presented in two parts. The first deals with theory, and techniques for applying that theory in the analysis of systems. A highlight of this (chapter five) is the development of modelling procedures that allow intractable periodic differential equations to be handled. This is regarded as significant since the great majority of differential equations with periodic coefficients cannot be treated by closed form methods of analysis. The second part presents an overview of the applications of periodic equations. The particular applications chosen have been done so to illustrate the variety of ways periodic differential equation descriptions arise and to demonstrate that the modelling procedures of Part I can be useful in determining system properties.

The developments in Part I and the applications of Part II are all related to a standard form for the equations, which in the case of second order systems is the canonical Hill equation used by McLachlan, viz.

$$\ddot{x} + (a - 2q\psi(t))x = 0, \psi(t) = \psi(t + \pi)$$

where $\psi(t)$ is a general periodic coefficient. Adopting such a canonical form is of value if the results of the theory and techniques chapters are to be used directly. Original system equations for particular applications presented in Part II are thus transformed into the appropriate canonical form before drawing upon the material of Part I.

The treatment is not intended to provide a text for the study of periodic differential equations but could be used for a single semester senior undergraduate or graduate level subject in systems with periodic parameters, particularly if applications are to be emphasised.

The encouragement and assistance of others is of course essential in producing a book. It is the author's pleasure to acknowledge the inspiration of his mentor and friend Professor Dick Huey to whom this book is dedicated; over the years of their association he has done much to encourage in a quiet, yet effective, manner the completion of this work. The manuscript was typed by Mrs. Gellisinda Galang to whom the author is grateful for the competent and patient manner in which she undertook the task.

Kensington, Australia, March 1982 J. A. Richards

Contents

List of Symbols

a	constant coefficient in a Hill or Mathieu equation
$a_0, a_1 \cdots$	stability boundaries for a Hill or Mathieu equation
A_{ir}	residues of the Hill determinant
$b_1, b_2 \cdots$	stability boundaries for a Hill or Mathieu equation
B_1, B_2	boundary modes for a Hill equation
c	capacitance per unit length
C	capacitance; discrete transition matrix; solution type of a Hill equation corresponding to complex eigenvalues of the discrete transition matrix
C_r	expansion coefficients in a Floquet solution
ce_m	Mathieu function of the first kind of order m
Ce_m	modified Mathieu function of the first kind of order m
ce_ν	Mathieu function of the first kind of fractional order ν
ceu_ν	unstable Mathieu function of the first kind of fractional order ν
$d_i(t)$	weighting coefficients
$\boldsymbol{D}$	electric displacement vector
e	eccentricity; charge on ion
E	Young's modulus
$\boldsymbol{E}$	electric field vector
$\boldsymbol{E}(t)$	time varying electric field
E_g	generator voltage
f	semi interfocal distance of an ellipse
$f(t), \boldsymbol{f}(t)$	forcing function, and vector form, in a periodic system; general function of time
F	force
$g_i(t)$	periodic coefficient in a general order Hill equation
$G(t)$	periodic coefficient matrix
G_n	complex Fourier coefficient of $g(t)$.
h	eigenvalue of a waveguide mode
h_k, h_r	step heights in a staircase periodic coefficient
$h(t)$	impulse response of a linear, time-invariant system
$\hbar$	$h/2\pi$, h is Planck's constant
$\boldsymbol{H}$	magnetic field vector
$\boldsymbol{i}, \boldsymbol{j}, \boldsymbol{k}$	unit vectors in a cartesian coordinate system
$i(t)$	current
I	current phasor; second moment of area
J_ν	Bessel function of the first kind of order ν
k	wave number
l	inductance per unit length
L	inductance
m	mass of an ion; distributed mass per unit length; waveguide mode number
M_+, M_-	state transition matrices
n	waveguide mode number
N	solution type of a Hill equation corresponding to a negative eigenvalue of the discrete transition matrix

$p(t)$	periodic staircase modulating function
P	solution type of a Hill equation corresponding to a positive eigenvalue of the discrete transition matrix
$P(t)$	periodic matrix
P_m	complex Fourier coefficient of $p(t)$
q	half amplitude of the periodic coefficient in a Hill or Mathieu equation
$q_{mn}^{c \text{ or } s}$	nth value of q that gives a zero of the mth order modified Mathieu function of the first or second kind
q	charge
$q(t)$	periodic staircase modulating function
Q_m	complex Fourier coefficient of $q(t)$
r	radial polar coordinate
r_0	field radius in a quadrupole mass filter
R	resistance
R_g	generator impedance
R_L	load impedance
R_s	series resistance of varactor diode
s	complex frequency variable; Laplace transform variable
s_i	root of system characteristic equation
se_m	Mathieu function of the second kind of order m
Se_m	modified Mathieu function of the second kind of order m
t	real time
T	period; tension
$T_r(x)$	Chebyshev polynomial
U	dc potential
v	phase velocity
$v(t)$	voltage
V	Voltage phasor; magnitude of periodic potential
$W(t)$	Wronskian matrix; unit periodic sampling function; impulse response of an N path network
W_0	Wronskian (determinant)
$\boldsymbol{x}(t)$	state vector for a periodically time-varying system
$\bar{X}(s)$	single sided Laplace transform of the state vector $\boldsymbol{x}(t)$
Y_ν	Bessel function of the second kind of order ν
α	real part of the characteristic exponent
α_k	constant coefficient in a general periodic differential equation
β	imaginary part of the characteristic exponent; phase constant inside a waveguide
β_0	phase constant in an unbounded medium
β_r	phase constant for the rth space harmonic in a periodic structure
γ	propagation constant inside a waveguide
γ_k	amplitude parameter in a general periodic differential equation
γ_0	propagation constant in an unbounded medium
Γ	exponent matrix associated with the discrete transition matrix; Gamma function
δ	duty cycle parameter for a rectangular or trapezoidal waveform
$\delta(t)$	Delta (impulse) function
$\Delta(\ )$	Hill infinite determinant
ε	permittivity; rise or fall time in a trapezoidal waveform
ζ	canonical displacement variable; damping constant
η	normalised rise and fall time in a trapezoidal waveform; elliptic coordinate
θ	period of the periodic coefficients in a general order Hill equation
λ_c	cut-off wavelength in a waveguide
λ_i	eigenvalue of the discrete transition matrix
Λ	matrix of eigenvalues of the discrete transition matrix
μ	permeability
μ, μ_i	characteristic exponent in the solution to a periodic differential equation

ξ	canonical time variable; elliptic coordinate
$\Xi(\ )$	periodic function in the solution to a periodic differential equation
π	pi; period of the periodic coefficient in a Hill or Mathieu equation
ρ_r	pole of the Hill determinant
σ	conductivity
τ	positive porch length in a rectangular or trapezoidal waveform
τ_r, τ_f	rise and fall times in a trapezoidal waveform
ϕ	scalar electric potential; polar coordinate
$\phi(t, 0)$	state transition matrix over the interval $(0, t)$
$\phi(\theta, 0)$	discrete state transition matrix for a general periodic differential equation
$\phi(\pi, 0)$	discrete state transition matrix for a Hill or Mathieu equation
χ	canonical time interval less than π
χ_{mn}^e	nth zero of the mth order Bessel function of the first kind
χ_{mn}^h	nth zero of the first derivative of the mth order Bessel function of the first kind
ψ	matrix of eigenvectors of the discrete transition matrix
$\psi(t)$	periodic function; voltage applied to the electrodes of a quadrupole mass filter
ψ_i	eigenvector of the discrete transition matrix
Ψ_n	complex Fourier coefficient of $\psi(t)$
ω	frequency
ω_c	cut-off frequency of a waveguide
ω_i	ith natural frequency of a linear, time-invariant system; system input (forcing) frequency
ω_l	idler frequency
ω_0	resonant frequency of a second order system
ω_p	pump frequency
∇^2	Laplacian in three dimensions
∇_t^2	transverse (two dimensional) Laplacian

Part I
Theory and Techniques

Chapter 1

Historical Perspective

1.1 The Nature of Systems with Periodically Time-Varying Parameters

The field of linear differential equations with *constant* coefficients has been extensively studied as a unified body of knowledge; standard forms of solution are well-known and Laplace transform techniques can be readily applied to obtain both natural and forced responses. Consequently a large proportion of all physical systems, including a majority of electrical network configurations, can be adequately described mathematically. However when the constraints of linearity and constant coefficients are relaxed, the neatness of solution is lost and very often particular non-linear and/or time-varying[1] systems and their associated differential equations have to be treated individually. Techniques devised for one type of system often cannot be generalised for use with another and consequently little or nothing is gained by developing stylised solution methods for the equations, such as those based upon integral transforms. Indeed it can even be difficult delineating classes of equation in many instances. A case which is an exception however is that of linear differential equations with coefficients that are periodically varying with time. As a class, so-called *periodic differential equations* exhibit similarities in behaviour, even though the solutions in most cases are not known in closed form, a feature which is exploited in Chapter five in developing modelling techniques for describing the dynamic behaviour of periodically varying systems.

Some systems with periodic parameters will lead to equations of quite high order, as could be encountered in electrical network problems. However, by far the most frequently met version of a periodic differential equation is of second order, expressible as

$$\ddot{x} + (a - 2q\psi(t))x = 0 \tag{1.1}$$

in which $\psi(t) = \psi(t + \pi)$. Thus $\psi(t)$ is a periodic coefficient—reflecting periodic parameter variations in a second order system—with a period chosen classically as π. Should a first derivative term be present it can be transformed out of the

[1] Throughout this book 'time-varying' is taken to include variations with other types of independent variable, such as displacement.

equation; (see Prob. 1.2). In Eq. (1.1), a represents a constant portion of the coefficient of x and q accounts for the magnitude of the time variation. In the particular case of q being zero, the equation reduces to an ordinary differential equation with constant coefficients (in this particular case often identified as the 'simple harmonic motion equation'). It is to be expected therefore that the solutions of time-varying equations for q small, will not be too different to simple sinusoids (or exponentials in the case of a negative).

Equation (1.1) is commonly known as Hill's equation and the form in which it is expressed is that most widely encountered.[2] When $\psi(t) = \cos 2t$ Eq. (1.1) becomes the Mathieu equation, perhaps the best known form of a Hill equation.

The rather unusual behaviour of a system described by a Hill equation (and, to an extent, by a periodic differential equation of order higher than two) can be appreciated by considering a very simple example from circuit theory. Straightforward as it is, this illustration is the basis for the useful technique of parametric amplification widely employed in satellite communication receivers and in radio telescopes.

The parallel LC network shown in Fig. 1.1 can be described by a second order equation in the capacitor voltage v; (see Prob. 1.1). If the capacitance is a periodic function of time then this equation will have some periodic coefficients and is thus a Hill equation. Imagine the capacitance can be varied in value by mechanically changing the plate separation and suppose energy has been introduced to the network at some earlier time so that it is now oscillating back and forth between the capacitance and the inductance at a rate determined by the resonant frequency of the combination, given by $\omega_0 = (LC)^{-1/2}$. This interchange of energy can be observed by examining the voltage across the capacitor, shown plotted in Fig. 1.2a.

At instants of time when the capacitor voltage is zero, all the energy of the circuit resides in the inductor whereas when the capacitor voltage is a maximum all the energy is stored in the capacitance. Suppose that when the capacitor voltage is a maximum (either positive or negative) the capacitor plates are suddenly pulled apart, thus decreasing capacitance instantaneously, as depicted in Fig. 1.2b. Work is done against the field between the plates of the capacitor and that work will be added to the overall energy of the network. At the particular instant of time under consideration—since all the energy is residing in the capacitance—the increase will be manifest as an increase, or amplification, of the capacitor voltage, as represented in Fig. 1.2c. Dynamic constraints prevent the voltage from taking a step increase as shown, nevertheless once it is increased, the capacitor voltage thereafter describes a new sinusoid of a larger amplitude than that before the capacitance change was invoked.

---

[2] Although not widespread, there are some variations in the symbols adopted for the coefficients. Instead of a and $-q$, some authors adopt η and γ; λ and $2h^2$; $4\omega^2$ and $4\alpha^2$; and R and $\pm 2h^2$.

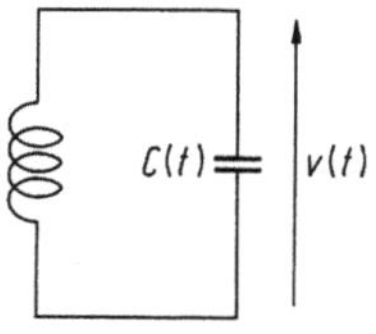

Fig. 1.1. Simple circuit with a time varying capacitance

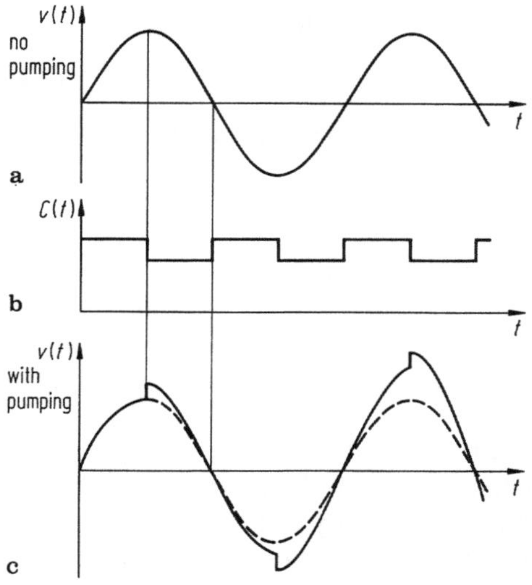

Fig. 1.2. a Unpumped capacitor voltage for the circuit of Fig. 1.1.; **b** Square wave capacitance variation; **c** Amplification of the capacitor voltage resulting from pumping the capacitor

When the capacitor voltage passes through zero the plates of the capacitor can be restored to their original position without affecting the energy in the circuit since, at that instant, all energy is stored in the inductor's magnetic field. This allows the plates to be separated again on the next capacitor voltage maximum, thereby increasing the network energy further. Clearly this process can be repeated at every maximum in the capacitor voltage, with the capacitance being restored at every voltage zero. As a consequence energy can be added to the network periodically leading to a continual amplification of the capacitor voltage (and, of course, all other network variables). When losses are present, for example by the inclusion of a series or shunt resistance, the unbounded amplification apparently available is limited, as it will also be in practice by the energy providing capacity of the mechanism which varies the capacitor.

The procedure of varying the capacitance in a periodic fashion to add energy to the circuit is referred to as 'pumping'. This is to maintain the distinction with 'forcing', which is the addition of energy directly from an input excitation such as a voltage or current source.

This illustration demonstrates that by pumping the capacitance at a rate equal to twice the frequency of the capacitor voltage—i.e. at twice the resonant frequency of the system—growing oscillations can be invoked in the lossless case whereas, with losses present, bounded amplification can be achieved. Even in the damped system growing oscillations can be produced if sufficient energy is added. Since

Fig. 1.3. Stretched string with periodically varying tension

unlimited growing oscillations infer instability, it is only necessary to pump the capacitance at twice the resonant frequency to induce instability in an otherwise stable, passive system. The condition for invoking instability therefore is

$$\omega_p = 2\omega_0 \tag{1.2}$$

where ω_p is the rate of pumping. It is shown in Chap. 4 that this can be generalised to

$$\omega_p = 2\omega_0/n \tag{1.3}$$

where n is an integer, although the damping effect of losses becomes more significant as n increases. As a result parametric[3] instability, invoked by pumping, is often only observed with $n = 1$. This implies also, in the case of a system stabilised by losses, that more parametric gain is available when pumping with $n = 1$ than with higher values.

In practice the capacitance in the illustration would not be pumped mechanically but rather an electrically variable capacitance would be employed, as described in Chap. 7. Also, notwithstanding the availability of means for achieving it, the effects observed in the network of Fig. 1.1 could just as well have been induced by pumping the inductance. Indeed, in general, parametric behaviour will be produced if any of a system's energy storing parameters are pumped. More specifically, parametric effects depend upon perturbing the natural frequency by pumping an energy storing parameter. For purposes of description, the unpumped natural frequency of a system is often referred to as its *static* natural frequency. Thus to ascertain whether a particular system will exhibit parametric behaviour it is only necessary to examine the expression for its static natural frequency and choose from that expression an energy storage parameter to be pumped. It is necessary to make this choice based upon inspection of the static natural frequency, since some energy storage parameters—notably the mass of the bob in a pendulum—do not contribute to the determination of the natural frequency and thus will not lead to system parametric behaviour if pumped. As an illustration of this principle consider the inducement of parametric oscillations on the stretched string shown in Fig. 1.3. Provided the displacements of the string are small, to ensure constant tension, the natural frequencies of vibration are given by (see A. H. Churches: Mechanical vibrations. New York: Wiley 1957)

---

[3] The phenomena are referred to as 'parametric' owing to their dependence upon a parameter variation.

$$\omega_0 = \frac{n\pi}{L}\sqrt{\frac{T}{m}} \tag{1.4}$$

where L is the length of the string, T its tension, m the distributed mass of the string and n is an integer. Consequently if any of T, m or L are varied at a rate of $2\omega_0$ growing oscillations should be observed on the string. Clearly m and L cannot be varied conveniently although the tension can, by varying the force F, along the string, as shown in the figure. This could be carried out using an electric motor with a suitable eccentric wheel attached to it, or even, as suggested by Rayleigh, by passing a pulsed electric current through the string if it were of a suitable conductor—such as steel. Clearly thermal time constants would need to be considered in the second approach.

If the tension in the string is varied at $2\omega_0$, then vibration will occur in the corresponding mode—i.e. for the particular value of n associated with the chosen ω_0. These vibrations will grow in amplitude until nonlinear restraints are encountered, in which case the linear parametric behaviour of the system is limited by the nonlinearities.

In a similar manner to the above illustration, conditions for parametrically induced behaviour of other physical systems can often be deduced readily from a simple inspection of their static properties. This principle will be of value in the following sections in which the history of parametric effects is reviewed, culminating in brief discussions on important present-day applications. Attention is given mainly to the effects themselves and the systems which give rise to them, whereas the history of the associated mathematics is not explicitly covered. This is traced very well to about 1940, in McLachlan's comprehensive book dealing with the Mathieu equation [1]. More recent mathematical developments are summarised in Sect. 1.5 to follow.

1.2 1831–1887 Faraday to Rayleigh—Early Experimentalists and Theorists

The first recorded demonstration of parametric behaviour appears to have been that of Faraday in 1831 [2] in which he produced wave motion in fluids, such as air, oil and water by vibrating a plate or membrane in contact with the fluid. The frequency of the waves so produced was one half of that of the vibrating plate, in agreement with Eq. (1.2) above. Faraday's experiments were carried out to provide a better explanation of the effects he and others observed than that given in 1827 by Savart. Faraday's conclusions however were apparently opposed by those of a Dr. L. Matthiessen in 1870, which prompted Lord Rayleigh to turn his attention to the interesting range of parametric behaviour [3]. Rayleigh refined and repeated Faraday's experiments and paid a deal of attention to means for measuring the frequency of pumping and the frequency of response, showing indeed that the relationship of Eq. (1.2) applied and that Faraday's explanations were plausible. Rayleigh also emphasised that waves can be parametrically excited in a wine glass, partially filled, by exciting the glass boundary in the well-known manner of drawing a moistened finger around the circumference. When the glass 'sings', fine wavelets, or 'crispations' as Rayleigh and Faraday termed them, will be observed to be formed on the surface of the fluid.

The inducement of vibrations in a stretched string by varying its tension periodically, as described above, was demonstrated in 1860 by Melde [4]. In his experiment the tension was pumped by attaching one end of the string to a vibrating tuning fork.

Although concerned with the natural modes of vibration of lakes with elliptical boundaries, and not parametric behaviour as such, the first detailed theory relevant to the study of periodically time-varying systems was given by Mathieu in 1868 and 1873 [5, 6]. The celebrated equation which now bears his name arose from these studies and, in an analogous manner, applies to the study of wave motion of all types subject to elliptical boundary constraints—the theory of elliptical waveguides considered in Sect. 7.3 of this book is a particular example.

Shortly afterwards, in 1883, Floquet [7], introduced what is commonly known as Floquet Theory—this is developed in more modern terminology in the next chapter. Floquet theory forms the basis of a great many of the descriptions of parametric behaviour, especially in spatially distributed systems in which *position* is an independent variable, such as in periodically loaded waveguides and transmission lines. Although Floquet theory is often quoted and used surprisingly little reference is ever made to that author's original publication.

One of the most significant and important of the earlier papers on the behaviour of periodically time-varying systems was that by Hill in 1886 [8], although it was preceded by an earlier publication on the topic by the same author in 1877. It was an important work in that it laid the very mathematical foundations of the stability theory of parametric systems. Hill developed an infinite determinant description of periodic systems. The determinants involved have, since that time, been referred to as Hill (infinite) determinants and are used extensively throughout this book both for stability studies and also in the justification of the modelling techniques for solution treated in Chap. 5. The significance of Hill's paper lies in the fact that it appears to have been the first investigation and solution of a practical problem in the field of periodically time-varying systems. To that time astronomers had been puzzled by the motion of the lunar perigee. Hill used a periodic differential equation description which gave a satisfactory explanation of the effect in terms of the varying gravitational influences of the sun and moon in motion.

As noted earlier, Hill's name has been given to the general class of second order periodic differential equations and his infinite determinants have been the subject of extensive mathematical treatments [9, 10].

Hill's paper was the seed for a further study by Lord Rayleigh in 1887 [11]. Rayleigh generalised Hill's mathematical derivations generating, for the first time, a theory for second order systems including losses or damping. In this paper Rayleigh also commented on the selective band pass nature of media whose transmission properties vary periodically in the direction of propagation. In particular he described selective transmission and reflection of light of different colours through glass which has become laminated through superficial decomposition, and he further suggested that similar dynamic effects could be observed in periodically loaded stretched strings. All of these illustrations of course are simply forerunners of many of the periodically loaded distributed systems encountered in present times, such as loaded waveguides and transmission lines.

1.3 1918–1940 The First Applications

There appears to have been little activity in the field of parametric systems, with regard to applications, for the thirty years to 1918, apart perhaps from Stephenson's studies [12, 13]. However this dearth of interest was more than compensated in the following twenty odd years. In 1918 a rather innocuous but very important paper, dealing with problems of instability in the side rods of locomotives, was published by Meissner [14]. This study led to an equation of the form of Eq. (1.1) in which $\psi(t)$ is a rectangular function. This particular form has become known as the Meissner equation and has the appeal that, unlike the Mathieu equation, it can be solved exactly. This is a consequence of the fact that it can be viewed as a pair of constant coefficient equations each valid in alternating time intervals. Shortly after Meissner's paper, Carson [15] published results of a study into frequency modulation based upon the Mathieu equation (it can be seen that the circuit of Fig. 1.1 earlier can be used as a frequency modulator simply by varying the capacitance in accord with the modulating signal). Jeffreys in 1924 [16] discussed some approximate methods for solving the Mathieu equation in the context of hydrodynamic problems and M. J. O. Strutt (as against J. W. Strutt—Lord Rayleigh) in 1927 [17] investigated the problems of eddy currents in elliptical conductors, a problem similar in principle to that studied by Mathieu. Strutt also devoted attention to the mathematical aspects of periodically time-varying systems in 1929 [18] and published a monograph in 1932 [19] dealing with applications of Mathieu and related functions. Mathieu functions are particular solutions of the Mathieu equation.

In 1928 a paper important to the essential philosophy of Chap. 5 of this book appeared. Van der Pol and Strutt [20] considered the problem of modelling the stability behaviour of the difficult-to-treat Mathieu equation by the readily deduced properties of the Meissner equation. Although no steps were taken to adjust the parameters of the Meissner equation to obtain the best approximation possible, theirs appear to have been the first departure from rigorous mathematical treatments of the Mathieu equation to techniques which, albeit approximate, are usable.

A celebrated paper in the field of quantum mechanics was published by Kronig and Penny in 1931 [21]. Based upon ideas generated earlier by Bloch, who suggested the periodic nature of the potential experienced by conduction electrons in crystals, Kronig and Penny solved the Schrödinger equation with a rectangular potential energy function. In this case the Schrödinger equation is simply of the form of Meissner's equation and thus can be solved exactly—this enables the exact calculation of allowed energies for electrons in crystals. Kronig and Penny also considered the limiting case of the Meissner equation in which the rectangular potential degenerates to a periodic sequence of impulses. Theirs appears to have been the first treatment of such a Hill equation.

In the 1930's several applications oriented publications appeared [22, 23], including the timely paper by Chu [24] dealing with elliptical waveguides. Again this is a restatement of Mathieu's original problem, with the waveguide modes appearing as Mathieu functions (as against trigonometric functions for rectangular guides and Bessel functions for circular waveguides).

1.4 Second Generation Applications

In about 1950 a new class of application appeared in which systems were *synthesised* to take advantage of the properties offered by periodically time-varying parameters. At about that time particle accelerators such as the cyclotron were gaining importance in the field of nuclear physics; however difficulties were being encountered with constraining the spread of the particle beams during acceleration. Static focussing with normal electromagnetic lens arrangements appeared limited in usefulness. However Christophilos[4] found that improved overall focussing could be obtained if the particle beams were subjected to alternating focussing and defocussing influences at the correct rate and in the correct proportions. In this manner the particles in motion through the accelerator experienced a periodic variation in constraining force, leading to descriptive equations of the Hill variety. It is shown in Chap. 2 that one form of solution to a Hill equation is a modulated sinusoid—i.e. one which, though oscillatory in general about a mean position, refocusses in a periodic manner near to its initial condition. Such a solution is shown in Fig. 1.4. Particles in transit through an accelerator, in which periodic focussing and defocussing is used, will follow tightly bound trajectories similar to that illustrated in Fig. 1.4 and thus will have their spatial dispersion constrained. The technique has come to be known as *strong focussing* and is used in such particle accelerators as the *synchrotron*. Courant, Livingston and Snyder [25] published details of the strong focussing mechanism of this machine in 1952.

In the following year the same principle was used to devise a mass spectrometer. Paul and Steinwedel [26] constructed a quadrupole electrode arrangement similar in principle to the strong focussing lens of a particle accelerator. Ions are fired through this structure and undergo equations of motion in the form of a Hill equation, the periodic coefficient arising because of a time-varying potential applied to the electrodes. In the device, known commonly as a *quadrupole mass spectrometer* the inducement of parametric *instability* (cf. Fig. 1.2) in the ion trajectories by small changes in operating parameters is used to remove ions of all mass but one from the sample being analysed. The remaining stable ions are then

---

[4] Apart from U.S. Patent Specifications in 1950 and 1951, and a brief mention in Scientific American, 188 (1953) 45–46, Christophilos' work was not published.

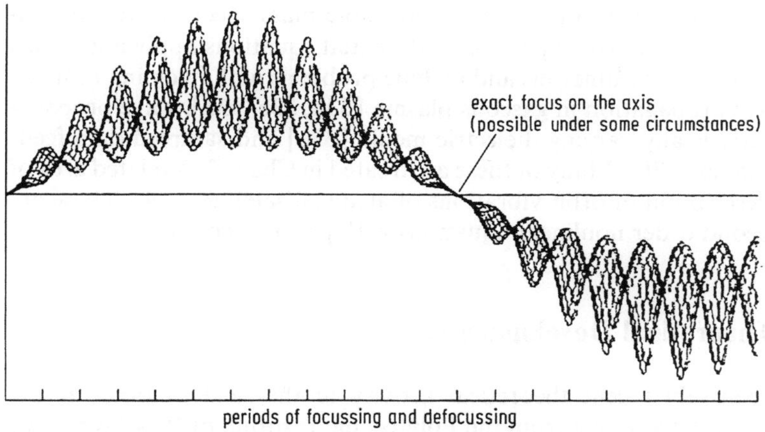

Fig. 1.4. Illustration of strong focussing of particle trajectories. A set of trajectories is shown, each corresponding to a different time of entry into the focussing field

counted to give a measurement of their concentration in the sample. This particular application of parametric principles has grown extensively since that time and has given rise to a number of new types of mass spectrometer. It is treated in depth in Chap. 7 and is also the subject of a recent book [27].

Owing to the ease of implementation with the availability of reliable multipliers, varactor diodes and the like, there have been a large number of electrical network configurations with periodically time-variable components examined since the early 1950's. Most of these depend upon stable, rather than unstable operation and take advantage of the frequency mixing offered when parametric systems are provided with an input, such as described in Sect. 3.14. Clearly in the electrical circuit case attention has not been restricted to second order. Whilst there are earlier references to networks with general component variations, the first comprehensive publications dealing with *periodic* component variations appear to be those of Smith [28], Desoer [29] and Fettweis [30]. However the most significant paper from an applications viewpoint would be that of Franks and Sandberg in 1960 [31]. This introduces the concept of the N-path network—a filter structure containing switches or modulators so that signal transmission can take place via N possible routes from input to output. N path networks are capable of a variety of circuit functions and are treated along with parametric amplifiers in Chap. 7. A very readable review of circuit-oriented parametric systems, dealing with papers to 1959, has been given by Mumford [32].

In 1953 Pipes published a key paper in practical methods for the analysis of systems described by Hill equations [33]. Apart from the earlier specific treatment of Van der Pol and Strutt [20], Pipe's publication is the first clear indication that intractable Hill equations are perhaps best treated by modelling them by tractable counterparts. In particular he suggested the use of a Hill equation with a periodic staircase-like coefficient wherein the coefficient is adjusted to resemble the shape of that of the difficult equation. Chapter 5 builds upon this concept to develop modelling techniques for Hill equations.

Apart from network applications and quadrupole mass spectrometers as discussed above, the fields to which periodic differential equations apply and which are of current interest are numerous and include problems with buckling in structures [33a], wave propagation in gaseous plasma [33b, 33c, 33d], wave propagation through periodically varying dielectric media [33e] and sampled data feedback control systems [33f]. Many of these are treated in Chap. 7. A related area of interest is the description of orbit vibrations of artificial satellites [34]. These are described by second order nonlinear equations with periodic coefficients.

1.5 Recent Theoretical Developments

Most of the relatively recent theoretical activity in the field of time-varying differential equations has come from the Polytechnic Institute of Brooklyn, New York [35–40, 40a] including a well-regarded monograph by Magnus and Winkler [41]. Loud [42], Trubowitz [42a], Mostaghel [42b], Berryman [42c], Taylor and Narrendra [43] and Gunderson, Rigas and van Vleck [44] have all contributed recent theoretical developments with the last two mentioned groups concentrating on the problem of predicting stability of the Mathieu equation with a first derivative, loss term present.

The treatments mentioned above have been confined to second order systems. By comparison the theoretical aspects of periodic differential equations of order high than two have also been studied. Perhaps the most recent papers dealing with high order equations and their specific properties would be those by Cooley, Clark and Buckner, who treat third order systems [45], Keenan, who also looks at third order parametric systems and their behaviour [46], and the general treatments by Sandberg [47], Meadows [48], Richards and Miller [49] and Richards and Cristaudo [50].

Of associated theoretical interest is the field of quasi-periodic or almost periodic systems. These lead to equations such as Eq. (1.1) but with almost periodic coefficients defined by

$$\psi(t) = \sum_n \Psi_n e^{j\omega_n t}$$

in which the ω_n are discrete but non-commensurate frequencies. Most of the earlier treatment of almost periodic functions and related differential equations can be found respectively in Bohr [50a] and Fink [50b]. Recent theoretical treatments include those of Dellwo and Friedman [50c], Davis and Rosenblat [50d] and Chow and Chiou [50e], the last of which deals strictly with randomly perturbed periodic coefficients and relates to systems of general order.

1.6 Commonplace Illustrations of Parametric Behaviour

The applications of periodic differential equations found today, as typified by those discussed in the preceding sections, range over a variety of fields and, in some cases, are highly sophisticated. Yet there is one particular parametric system,

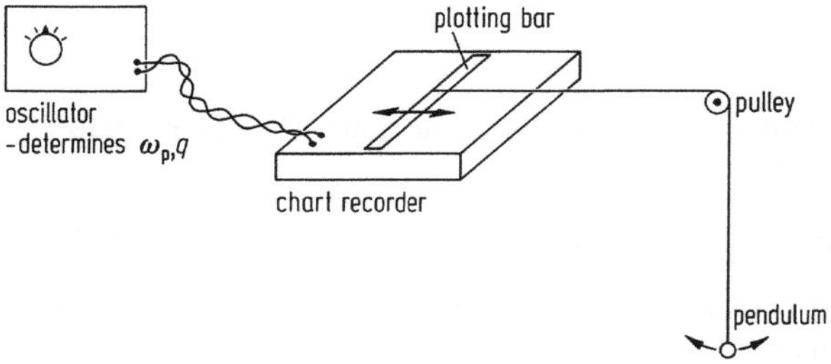

Fig. 1.5. Laboratory demonstration of parametric instability

operating according to the principles described in Sect. 1.1 above that is observed very commonly and by people in all walks of life. Indeed the properties of parametric behaviour are exploited most effectively in this system to the great enjoyment of children. This is simply the child riding on a playground swing. Technically the child on a swing can be viewed simply as a pendulum with a length which is variable, according to how the child seats himself on the swing and according also to how he holds his legs. Provided the child has an initial condition in the form of some displacement or velocity, he can increase his swinging motion by raising and lowering his legs synchronously with the swing's motion thereby periodically varying (i.e. 'pumping') the effective length of the pendulum and thus its natural frequency. The energy for the pumping here, of course, is internal energy 'stored' in the child. He can either pump at twice the swing frequency by lowering his centre of gravity towards each swing extremity and raising it when the swing passes through the vertical position, or else pump *at* the swing frequency by varying the effective pendulum length only once per period. The former situation is analogous to that of the circuit example of Sect. 1.1 above and is governed by Eq. (1.2) whereas the latter, which is the situation normally encountered in the playground, is a demonstration of parametric excitation according to Eq. (1.3) with $n = 2$. Figure 1.2 shows that pumping must be carried out with the correct phase relationship to achieve parametric gain—whether the system be a circuit or a swing. If the system is pumped at an incorrect phase initially, the natural response will adjust to maximise energy transfer from the pump source to the system to allow gain. This phase locking condition has been described by Keenan [51].

A very simple laboratory demonstration of the principles of parametric behaviour can be constructed, as shown in Fig. 1.5, using a pendulum. The length of the pendulum can be varied periodically by allowing the plotting bar of the $X - Y$ recorder to move in response to a signal derived from the oscillator. Control of the oscillator frequency determines the rate of pumping whereas control of the oscillator output determines the magnitude of pumping. By choosing an oscillator frequency equal to twice the static frequency of the pendulum, parametric amplification (tending to instability) can be quite convincingly demonstrated. Detuning the oscillator slightly will illustrate the sensitivity of the system response to having the pumping frequency precisely set. With care, and some little difficulty, other modes of parametric excitation, corresponding to $n \geq 2$ in Eq. (1.3), can be invoked. Moreover phase-locking of the natural response is easily observed by starting the pendulum swinging at a phase randomly timed with respect to the pump.

References for Chapter 1

1. McLachlan, N. W.: Theory and application of Mathieu functions. Clarendon: Oxford, U. P. 1947. Reprinted by Dover, New York 1964
2. Faraday, M.: On a peculiar class of acoustical figures; and on certain forms assumed by groups of particles upon vibrating elastic surfaces. Philos. Trans. R. Soc. London, 121 (1831) 229–318
3. Lord Rayleigh (Strutt, J. W.): On the crispations of fluid resting upon a vibrating support. Philos. Mag. 16 (1883) 50–58
4. Melde, F.: Über die Erregung stehender Wellen eines fadenförmigen Körpers. Ann. Phys. Chem. (Ser. 2) 109 (1860) 193–215
5. Mathieu, E.: Mémoire sur le mouvement vibratoire d'une membrane de forme elliptique. J. Math. Pure Appl. 13 (1868) 137
6. Mathieu, E.: Cours de mathématique physique. Paris 1873
7. Floquet, G.: Sur les équations differentiales linéaires. Ann. L'Ecole Normale Super. 12 (1883) 47
8. Hill, G. W.: On the part of the moon's motion which is a function of the mean motions of the sun and the moon. Acta Math. 8 (1886) 1–36
9. (a) von Koch, H.: Sur les déterminants infinis et les équations differentiales linéaires. Acta Math. 16 (1892/93) 217–295
   (b) —: Sur une application des déterminants infinis à la théorie des équations différentiales linéaires. Acta Math. 15 (1891) 53–63
   (c) —: Sur quelques points de la théorie des déterminants infinis. Acta Math. 24 (1901) 89–122
10. Forsyth. A. R.: Theory of differential equations, Part II, Vol. III. New York: Dover 1959
11. Lord Rayleigh (Strutt, J. W.): On the maintenance of vibrations by forces of double frequency and on the propagation of waves through a medium endowed with a periodic structure. Philos. Mag. (Ser. 5) 24 (1887) 143–159
12. Stephenson, A.: A class of forced oscillations. Q. J. Math. 37 (1906) 353
13. Stephenson, A.: New type of dynamical stability. Proc. Manch. Philos. Soc. 52 (1908)
14. Meissner, E.: Über Schüttelerscheinungen im System mit periodisch veränderlicher Elastizität. Schweiz. Bauztg. 72 (1918) 95–98
15. Carson, J. R.: Notes on theory of modulation. Proc. IRE 10 (1922) 62
16. Jeffreys, H.: Approximate solutions of linear differential equations of second order. Proc. Lon. Math. Soc. 23 (1924) 428
17. Strutt, M. J. O.: Wirbelströme im elliptischen Zylinder. Ann. Phys. 84 (1927) 485
18. Strutt, M. J. O.: Der charakteristische Exponent der Hillschen Differentialgleichung. Math. Ann. 101 (1929) 559–569
19. Strutt, M. J. O.: Lamésche, Mathieusche und verwandte Funktionen in Physik und Technik. 1932
20. van der Pol, B.; Strutt, M. J. O.: On the stability of solutions of Mathieu's equation. Philos. Mag. 5 (1928) 18–39
21. de Kronig, R. L.; Penney, W. G.: Quantum mechanics of electrons in crystal lattices. Proc. R. Soc. (Ser. A) 130 (1931) 499–513
22. (a) Erdelyi, A.: Über die freien Schwingungen in Kondensatorkreisen mit periodisch veränderlicher Kapazität. Ann. Phys. 19 (1934) 585
    (b) —: Zur Theorie des Pendelrückkopplers. Ann. Phys. 23 (1935) 21
23. Barrow, W. L.; Smith, D. B.; Baumann, F. W.: Oscillatory circuits having periodically varying parameters. J. Frank. Inst. 221 (1936) 403
24. Chu, L. H.: Electromagnetic waves in elliptical metal pipes. J. Appl. Phys. 9 (1938) 583
25. Courant, E. D.; Livingston, M. S.; Snyder, H. S.: The strong focussing synchrotron—A new high energy accelerator. Phys. Rev. 88 (1952) 1190–1196
26. Paul, W.; Steinwedel, H.: Ein neues Massenspektrometer ohne Magnetfeld. Z. Naturforsch. 8a (1953) 448–450

27. Dawson, P. H.: Quadrupole mass spectrometry. Amsterdam: Elsevier 1976
28. Smith, B. D.: Analysis of commutated networks. IRE Trans. Aeronaut. Navig. Electron. AE-10 (1953) 21–26
29. Desoer, C. A.: Steady-state transmission through a network containing a single time-varying element. IRE Trans. Circuit Theory CT-6 (1959) 244–252
30. Fettweis, A.: Steady-state analysis of circuits containing a periodically-operated switch. IRE Trans. Circuit Theory CT-6 (1959) 252–260
31. Franks, L. E.; Sandberg, I. W.: An alternative approach to the realization of network transfer functions: the *N*-path filter. Bell Syst. Tech. J. 39 (1960) 1321–1350
32. Mumford, W. M.: Some notes on the history of parametric transducers. Proc. IRE 48 (1960) 848–853
33. Pipes, L. A.: Matrix solution of equations of the Mathieu-Hill type. J. Appl. Phys. 24 (1953) 902–910
33a. Timoshenko, S. P.; Gere, J. M.: Theory of elastic stability, 2nd ed., New York: McGraw-Hill 1961
33b. Lehane, J. A.; Paoloni, F. J.: Parametric amplification of Alfven waves. Plasma Phys. 14 (1972) 461–471
33c. Cramer, N. F.: Parametric excitation of ion-cyclotron waves. Plasma Phys. 17 (1975) 967–972
33d. Cramer, N. F.; Donelly I: Parametric excitation of kinetic Alfven waves. Plasma Phys. 26 (1981) 253
33e. Elachi, C: Waves in active and passive periodic structures: A review. Proc. IEEE 64 (1976) 1666–1698
33f. Hiller, J; Keenan, R. K.: Stability of finite width sampled data systems. Int. J. Control 8 (1968) 1–22
34. Anand, D. K.; Yuhasz, R. S.; Whisnant, J. M.: Attitude motion in an eccentric orbit. J. Spacecr. Rockets 8 (1971) 903–905
35. Levy, D. M.; Keller, J. B.: Instability intervals of Hill's equation. Comm. Pure Appl. Math. 16 (1963) 469–476
36. Hochstadt, H.: A special Hill's equation with discontinuous coefficients. Am. Math. Monthly 70 (1963) 18–26
37. Hochstadt, H.: Instability intervals of Hill's equation. Comm. Pure Appl. Math. 17 (1964) 251–255
38. Hochstadt, H.: On the stability of certain second order differential equations. J. Soc. Ind. Appl. Math. 12 (1964) 58–59
39. Hochstadt, H.: A stability estimate for differential equations with periodic coefficients. Arh. Math. 15 (1964) 318–320
40. Hochstadt, H.: An inverse problem for Hill's equation. J. Diff. Eq. 20 (1976) 53–60
40a. Goldberg, W; Hochstadt, H; On a Hill's equation with selected gaps in its spectrum. J. Diff. Eqs. 34 (1979) 167–178
41. Magnus, W.; Winkler, S.: Hill's equation. New York: Wiley 1966
42. Loud, W. S.: Stability regions for Hill's equation. J. Diff. Eq. 19 (1975) 226–241
42a. Trubowitz, E: The inverse problem for periodic potentials. Comm. Pure Appl. Math. XXX (1977) 321–337
42b. Mostaghel, N: Stability regions of Hill's equation J. Inst. Math. Appl. 19 (1977) 253–259
42c. Berryman, J. G.: Floquet exponent for instability intervals of Hill's equation. Comm. Pure Appl. Math. XXXll (1979) 113–120.
43. Taylor, J. H.; Narendra, K. S.: Stability regions for the damped Mathieu equation. SIAM J. Appl. Math. 17 (1969) 343–352
44. Gunderson, H.; Rigas, H.; van Vleck, F. S.: A technique for determining the stability of the damped Mathieu equation. SIAM J. Appl. Math. 20 (1974) 345–349
45. Cooley, W. W.; Clark, R. N.; Buckner, R. C.: Stability in a linear system having a time-variable parameter. IEEE Trans. Autom. Control AC-9 (1964) 426–434
46. Keenan, R. K.: Exact results for a parametrically phase-locked oscillator. IEEE Trans. Circuit Theory CT-14 (1967) 319–335

47. Sandberg, I. W.: On the stability of solutions of linear differential equations with periodic coefficients. J. Soc. Ind. Appl. Math. 12 (1964) 487–496
48. Meadows, H. E.: Solutions of systems of linear ordinary differential equations with periodic coefficients. Bell Syst. Tech. J. 41 (1962) 1276–1294
49. Richards, J. A.; Miller, D. J.: Features of mode diagrams for lth order periodic systems. SIAM J. Appl. Math. 25 (1973) 72–82
50. Richards, J. A.; Cristaudo, P. G.: Parametric aspects of mode and stability diagrams for general periodic systems. IEEE Trans. Circuit Syst. CAS-24 (1977) 241–247
50a. Bohr, H: Almost periodic functions. (transl. Cohn, H; Steinhardt, F.) New York: Chelsea 1947
50b. Fink, A. M.: Almost periodic differential equations. Berlin, Heidelberg, New York: Springer 1974
50c. Dellwo, D. R.; Friedman, M. B.: Uniform asymptotic solutions for a differential equation with an almost periodic coefficient. SIAM J. Appl. Math. 36 (1979) 137–147
50d. Davis, S. H.; Rosenblat, S.: A quasi-periodic Mathieu-Hill equation. SIAM J. Appl. Math. 38 (1980) 139–155
50e. Chow, P. L.; Chiou, K. L.: Asymptotic stability of randomly perturbed linear periodic systems. SIAM J. Appl. Math. 40 (1981) 315–326
51. Keenan, R. K.: An investigation of some problems in periodically parametric systems. PhD Thesis, Monash University Melbourne, 1966

Problems

1.1 For the circuit of Fig. 1.1 the inductor is described by $v(t) = L\dfrac{\mathrm{d}i(t)}{\mathrm{d}t}$ and the capacitor by $q(t) = C(t)v(t)$. Noting that $i(t) = \dfrac{\mathrm{d}q(t)}{\mathrm{d}t}$ derive differential equations in terms of $q(t)$ and $v(t)$. For $C(t)$ periodic, note that the charge equation is directly of the form of Eq. (1.1) but that the voltage equation contains a first derivative term.

1.2 Show that the transformation

$$v(t) = \exp\left\{-\tfrac{1}{2}\int_0^t g_1(t)\mathrm{d}t\right\}y(t)$$

changes the voltage equation in Prob. 1.1 to the form

$$\ddot{y} + \{g_0(t) - \tfrac{1}{2}\dot{g}_1(t) - \tfrac{1}{4}g_1(t)^2\}y = 0$$

where $g_0(t)$ is the coefficient of $v(t)$ and $g_1(t)$ is the coefficient of $\dfrac{\mathrm{d}v(t)}{\mathrm{d}t}$.

Suppose $C(t) = C_1 \cos \omega t$. Show that the equation in $y(t)$ is the same Hill equation as that for the charge in Prob. 1.1.

1.3 Suppose $C(t)$ is a general periodic capacitance. By the technique of Prob. 1.2 show that the charge and voltage equations for Fig. 1.1 can be reduced to the same Hill equation.

1.4 By deducing the equation for a pendulum under small oscillations determine the phase with which a child must move his legs in order to increase the oscillations of a swing. Does he normally invoke gain for $n = 1$ or $n = 2$ in Eq. (1.3)?

1.5 In principle the network of Fig. 1.1 could be used as a frequency modulator, yet instability in frequency modulators is not normally observed. Explain this by reference to Eq. (1.3).

1.6 Show that the motion of a particle moving in an electric field which changes in magnitude periodically with time is described by a differential equation with periodic coefficients.

Chapter 2

The Equations and Their Properties

Differential equations with periodically varying coefficients appear in a number of forms, some of which are known by specific names. In this chapter these forms are identified and the general conventions adopted in this book are outlined. Whilst some are expanded upon in later chapters, the essential properties of the equations are also summarised, particularly those of importance in practical applications.

2.1 Hill Equations

The most general periodic equation to be considered herein is of the form

$$g_v(t)\overset{v\cdot}{x} + g_{v-1}(t)\overset{(v-1)\cdot}{x} + \cdots + g_0(t)x = f(t) \tag{2.1}$$

in which the notation $\overset{v\cdot}{x}$ implies the vth derivative of x with respect to the independent variable t, which is usually time. The coefficients $g_i(t)$ may be constant or periodically varying with t. In the latter case it is assumed that they are frequency coherent and have a period θ. Thus

$$g_i(t) = g_i(t + \theta). \tag{2.2}$$

The function $f(t)$, generally associated with an impressed excitation, may also involve periodically varying coefficients. When $f(t)$ is zero Eq. (2.1) is said to be *homogeneous*, otherwise the equation is called *inhomogeneous*. The major portion of the theory of periodic differential equations is addressed to the homogeneous form since treatment of the inhomogeneous equation generally then follows by standard techniques.

The most commonly occuring periodic equations are of second order. Similarly the theory of second order equations is extensive with relatively little theory having been developed for particular higher order forms. The general second order equation is

$$\ddot{x} + g_1(t)\dot{x} + g_0(t)x = 0. \tag{2.3}$$

However the transformation

$$x(t) = \exp\left\{-\tfrac{1}{2}\int_0^t g_1(t)\,\mathrm{d}t\right\} y(t) \tag{2.4}$$

converts Eq. (2.3) into the following expression, without a first derivative,

$$\ddot{y} + \{g_0(t) - \tfrac{1}{2}\dot{g}_1(t) - \tfrac{1}{4}g_1(t)^2\}y = 0$$

Equation (2.4) can be used in a similar manner to remove the $(v - 1)$th derivative from Eq. (2.1) although with the $\frac{1}{2}$ replaced by v^{-1}. In its classical form the last expression is written as

$$\ddot{y} + (a - 2q\psi(t))y = 0, \quad \psi(t) = \psi(t + \pi) \tag{2.5}$$

which is known as the *Hill equation.* The function $\psi(t)$ has, by convention, a period of π and also obeys $|\psi(t)_{max}| = 1$; a is a constant parameter and $2q$ is a parameter which accounts for the magnitude of the time variation. Whilst the negative sign in the coefficient of Eq. (2.5) and the period of π can lead to some confusion in practice, if particular care is not taken, they have been retained in this treatment to maintain consistency with important discourses such as that by McLachlan [1].

The most widely known form of the Hill equation is the *Mathieu equation* in which $\psi(t)$ is sinusoidal:

$$\ddot{y} + (a - 2q\cos 2t)y = 0. \tag{2.6}$$

2.2 Matrix Formulation of Hill Equations

Equation (2.1) is perhaps most easily handled in application when it is expressed in matrix notation, an expedient often adopted in general for the computational solution of differential equations. By defining the column vector

$$\boldsymbol{x}(t) = \begin{bmatrix} x(t) \\ \dot{x}(t) \\ \vdots \\ \overset{(v-1).}{x(t)} \end{bmatrix}$$

referred to as *the state vector*, Eq. (2.1) can be recast as

$$\dot{\boldsymbol{x}}(t) = G(t)\boldsymbol{x}(t) + \boldsymbol{f}(t) \tag{2.7}$$

where $\boldsymbol{f}(t)$ is a vector describing the effect of the forcing function $f(t)$ in Eq (2.1). $G(t)$ is a matrix of constant or periodically varying coefficients. Again when $\boldsymbol{f}(t) = 0$ Eq. (2.7) becomes the homogeneous set of v first order periodic equations

$$\dot{\boldsymbol{x}}(t) = G(t)\boldsymbol{x}(t), \quad G(t) = G(t + \theta). \tag{2.8}$$

If $x_i(t)$ is a fundamental or basis solution to the homogeneous form of Eq. (2.1)—of which there are v linearly independent such solutions—then the corresponding fundamental vector solution of Eq. (2.8) is

$$\boldsymbol{x}_i(t) = \begin{bmatrix} x_i(t) \\ \dot{x}_i(t) \\ \vdots \\ \overset{(v-1).}{x_i(t)} \end{bmatrix}$$

There will be v of these fundamental vectors, the collection of which is called the Wronskian matrix

$$W(t) = [\boldsymbol{x}_1(t), \ldots \boldsymbol{x}_v(t)]$$
$$= \begin{bmatrix} x_1(t) & \cdots & x_v(t) \\ \dot{x}_1(t) & & \dot{x}_v(t) \\ \vdots & & \vdots \\ \overset{(v-1).}{x_1}(t) & & \overset{(v-1).}{x_v}(t) \end{bmatrix}. \tag{2.9}$$

The determinant of the Wronskian matrix is referred to simply as the *Wronskian*. It is a nonzero constant if the basis solutions $x_i(t)$ chosen for a particular equation do form a linearly independent set.

The technique of *variation of parameters* can be used to generate a form for the solution of Eq. (2.7). For this a general solution is assumed, of the type

$$\boldsymbol{x}(t) = \sum_{i=1}^{v} d_i(t)\boldsymbol{x}_i(t) = W(t)\boldsymbol{d}(t) \tag{2.10}$$

where the $d_i(t)$ are differentiable weighting functions and $\boldsymbol{d}(t)$ is a column vector of the $d_i(t)$'s. Substituting Eq. (2.10) into Eq. (2.7) leads to

$$\sum_{i=1}^{v} \dot{d}_i(t)\boldsymbol{x}_i(t) = \boldsymbol{f}(t)$$

or

$$W(t)\dot{\boldsymbol{d}}(t) = \boldsymbol{f}(t)$$

since the $\boldsymbol{x}_i(t)$ are solutions to Eq. (2.8) whereupon

$$\dot{\boldsymbol{d}}(t) = W(t)^{-1}\boldsymbol{f}(t)$$

so that upon integrating

$$\boldsymbol{d}(t) = \boldsymbol{d}(0) + \int_0^t W^{-1}(\tau)\boldsymbol{f}(\tau)\,\mathrm{d}\tau$$

which with Eq. (2.10) becomes

$$\boldsymbol{x}(t) = W(t)W^{-1}(0)\boldsymbol{x}(0) + \int_0^t W(t)W^{-1}(\tau)\boldsymbol{f}(\tau)\,\mathrm{d}\tau \tag{2.11}$$

which is the required general solution.

It is seen in the above that the Wronskian must be nonsingular, the condition noted earlier for the linear independence of the $x_i(t)$.

2.3 The State Transition Matrix

In Eq. (2.11) it is usual to write

$$\phi(t, 0) = W(t)W^{-1}(0). \tag{2.12}$$

$\phi(t, 0)$ is referred to as the *state transition matrix* over the interval $(0, t)$. Similarly $\phi(t, \tau) = W(t)W^{-1}(\tau)$, so that Eq. (2.11) can be rewritten as

$$\boldsymbol{x}(t) = \phi(t, 0)\boldsymbol{x}(0) + \int_0^t \phi(t, \tau)\boldsymbol{f}(\tau)d\tau. \tag{2.13}$$

Clearly once the transition matrices have been determined the complete solution to Eq. (2.1) is known. The first righthand term in Eq. (2.13) is the complementary function part of the complete solution whilst the second term is the particular integral. Alternatively these are often referred to as the *zero-input state response* and *zero-state state response* respectively.

At important property of the transition matrix $\phi(t, 0)$ is that it satisfies the homogeneous equation. This is seen by differentiating the complementary function to give

$$\dot{\boldsymbol{x}}(t) = \dot{\phi}(t, 0)\boldsymbol{x}(0).$$

With Eq. (2.8) this yields

$$\dot{\phi}(t, 0)\boldsymbol{x}(0) = G(t)x(t) = G(t)\phi(t, 0)\boldsymbol{x}(0)$$

which requires

$$\dot{\phi}(t, 0) = G(t)\phi(t, 0) \qquad \text{q.e.d.}$$

Similarly

$$\begin{aligned}\dot{\phi}(t + \theta, 0) &= G(t + \theta)\phi(t + \theta, 0)\\ &= G(t)\phi(t + \theta, 0) \text{ for } G(t) \text{ periodic;}\end{aligned}$$

therefore $\phi(t + \theta, 0)$ also satisfies the homogeneous equation.

2.4 Floquet Theory

Matrices which satisfy the homogeneous equation are referred to as fundamental matrices. Two fundamental matrices for the same equation can only differ by a nonsingular constant matrix. Thus

$$\phi(t + \theta, 0) = \phi(t, 0)C \tag{2.14a}$$

and in particular

$$\phi(\theta, 0) = \phi(0, 0)C. \tag{2.14b}$$

However from the complementary function of Eq. (2.13)

$$\boldsymbol{x}(0) = \phi(0, 0)\boldsymbol{x}(0)$$

so that $\phi(0, 0) = \boldsymbol{I}$, whereupon Eq. (2.14b) yields

$$C = \phi(\theta, 0). \tag{2.15}$$

C is referred to as the *discrete transition matrix*, describing system behaviour over one full period of the coefficients.

Equations (2.14a) and (2.15) together show

$$\phi(m\theta, 0) = \phi(\theta, 0)^m, \quad m\text{-integral}. \tag{2.16}$$

Moreover

$$\boldsymbol{x}(t + \theta) = C\boldsymbol{x}(t), \tag{2.17a}$$

whilst in general

$$\boldsymbol{x}(t + m\theta) = C^m\boldsymbol{x}(t). \tag{2.17b}$$

Equations (2.17) are most significant in the study of periodic differential equations since they infer that if a solution is known over one full period of the time variations in the system then the solution is known for all time. Equation (2.17a) is a vector statement of the celebrated *Floquet Theorem* which says that the solution to a periodic equation is related to the solution one full period away by a complex constant; i.e., in algebraic form,

$$x(t + \theta) = \sigma x(t).$$

It is convenient to put

$$C = \exp(\Gamma\theta), \text{ where } \Gamma \text{ is a matrix}, \tag{2.18}$$

and to define

$$P(t) = \phi(t, 0)\exp(-\Gamma t). \tag{2.19}$$

Thus

$$\begin{aligned} P(t + \theta) &= \phi(t + \theta, 0)\exp[-\Gamma(t + \theta)] \\ &= \phi(t, 0)C\exp(-\Gamma t)\exp(-\Gamma\theta) \\ &= P(t), \end{aligned}$$

i.e. $P(t)$ is periodic with period θ. From Eq. (2.19)

$$\phi(t, 0) = \exp(\Gamma t)P(t)$$

so that the unforced response of a periodic system can be written

$$\boldsymbol{x}(t) = \phi(t, 0)\boldsymbol{x}(0) = \exp(\Gamma t)P(t)\boldsymbol{x}(0) \tag{2.20}$$

which is a vector counterpart of a result well known from the algebraic treatment of periodic differential equations, viz.

$$x(t) = \sum_{i=1}^{v} \exp(\mu_i t)\Xi_i(t). \tag{2.21}$$

In the last expression the $\Xi_i(t)$ are bounded and periodic with period θ. The μ_i are complex constants and are referred to as *characteristic exponents*. The characteristic exponents are just the eigenvalues of the matrix Γ, whilst the eigenvalues of the discrete transition matrix $C = \phi(\theta, 0)$ are sometimes called *characteristic multipliers*. Equation (2.18) shows that the characteristic multipliers λ_i are related to the characteristic exponents according to

$$\lambda_i = \exp(\mu_i\theta). \tag{2.22}$$

A relationship of practical importance is

$$\prod_{i=1}^{v} \lambda_i = \exp\left\{\int_0^\theta \operatorname{trace}[G(t)]\mathrm{d}t\right\} \tag{2.23a}$$

$$= \exp\{-\langle g_{v-1}(t)\rangle\theta\} \tag{2.23b}$$

$$= \det\{\phi(\theta, 0)\}. \tag{2.23c}$$

In Eq. (2.23b) $\langle g_{v-1}(t)\rangle$ is the average value of $g_{v-1}(t)$, over a period θ.
It is important to realise that the v entries in Eq. (2.21) are the v linearly independent basis solutions. Should a number of μ_i be degenerate, t-multiplied forms are adopted, in the usual manner.

2.5 Second Order Systems

For second order, lossless systems as described by Eq. (2.5)

$$G(t) = \begin{bmatrix} 0 & 1 \\ -(a - 2q\psi(t)) & 0 \end{bmatrix}$$

so that Eq. (2.23a) gives

$$\prod_{i=1}^{2} \lambda_i = \lambda_1\lambda_2 = 1. \tag{2.24a}$$

Equation (2.22) therefore yields

$$\begin{aligned} \mu_1 + \mu_2 &= 0 \\ \mu_2 = -\mu_1 &= -\mu. \end{aligned} \tag{2.24b}$$

Equation (2.21) thus becomes

$$x(t) = \exp(\mu t)\Xi(t) + \exp(-\mu t)\Xi(-t). \tag{2.25}$$

The characteristic equation of $\phi(\theta, 0)$, used for determining the λ_i, is easily shown to be

$$\lambda^2 - \operatorname{trace}\{\phi(\theta, 0)\}\lambda + \det\{\phi(\theta, 0)\} = 0 \tag{2.26a}$$

which in view of Eqs. (2.23c) and (2.24a) becomes

$$\lambda^2 - \operatorname{trace}\{\phi(\theta, 0)\}\lambda + 1 = 0. \tag{2.26b}$$

The solutions of Eq. (2.26b) are

$$\lambda_{1,2} = T/2 \pm \sqrt{(T/2)^2 - 1} \tag{2.27a}$$

with $T = \operatorname{trace}\{\phi(\theta, 0)\}$. Thus, using Eq. (2.22)

$$\lambda_1 + \lambda_2 = e^{\mu_1\theta} + e^{\mu_2\theta} = T.$$

In view of Eq. (2.24b), the last expression gives

$$\cosh \mu\theta = T/2$$

so that

$$\mu = \frac{1}{\theta}\cosh^{-1}\left[\frac{\text{trace}\,\{\phi(\theta, 0)\}}{2}\right]. \tag{2.27b}$$

Equation (2.27b) demonstrates how μ can be determined computationally in a very convenient manner, once the discrete transition matrix for a Hill equation is known.

Second order systems with losses lead to an equation with a first derivative. This is easily handled using the transformation of Eq. (2.4), and then adopting the techniques just outlined. Alternatively Eqs. (2.23) can be used directly to give

$$\lambda_1\lambda_2 = \exp\{-\langle g_1(t)\rangle\theta\} \tag{2.28}$$

and $\mu_1 + \mu_2 = -\langle g_1(t)\rangle$, in general for an equation with a first derivative. Eq. (2.26a) still holds.

2.6 Natural Modes of Solution

It is clear from Eqs. (2.21) and (2.22) that each basis solution is associated with an eigenvalue of the discrete transition matrix so that the form of the basis solution is dependent upon the nature of eigenvalue. For example for a real positive or negative eigenvalue the associated basis solution will assume an essential form similar to that of the appropriate $\Xi_i(t)$, but with an increasing or decreasing exponential modulation. Accordingly it is useful to classify the types of natural solution according to the eigenvalues. Since eigenvalues can be positive, negative and complex the corresponding solutions are referred to as P type, N type and C type respectively.

Equation (2.22) shows that the characteristic exponent for a P type solution must be real so that the form of that solution is an exponentially modulated periodic function, of period θ, as noted above. Whether the modulation is increasing or decreasing clearly depends upon the value of μ, however Eq. (2.24b) demonstrates that one of each type of P solution will exist for a lossless second order system, if the eigenvalues of the discrete transition matrix are positive.

For an N type solution the characteristic exponent is seen from Eq. (2.22) to be of the form

$$\mu = \alpha + \mathrm{j}(2n + 1)\frac{\pi}{\theta}, \quad n\text{-integer}.$$

Usually the choice $n = 0$ is made so that the associated basis solution is

$$e^{\alpha t} e^{j\pi t/\theta} \Xi(t).$$

Again this is seen to be an exponentially modulated periodic function, with the direction of the modulation dependent upon the sign of α. However the period of the periodic term is now 2θ—i.e. it is a subharmonic of the time variation. The same properties will be obtained should other values of n be chosen in the above expression for μ.

A C type solution has a characteristic exponent of the general form

$$\mu = \alpha + j\beta$$

and thus will appear, in general, as an exponential modulation of the product of two periodic functions. (This will be recognised as exponentially modulated double sideband suppressed carrier behaviour.)

Note that for a second order lossless system $\alpha = 0$ for a C solution since $|\lambda| = 1$.

The collection of P, N and C types solutions for a given set of parameter values is called a *mode* of the system. It is a straightforward matter to show for a second order lossless system that the only allowable modes are $2P$, $2N$ and $2C$, implying that for a given set of values the mode of the system is two P type solutions, and so on.

There are a secondary set of modes, sometimes referred to as Brillouin modes, describing the system with degenerate sets of eigenvalues. These are discussed in Chap. 4 in a fuller treatment of system modes.

2.7 Concluding Comments

The foregoing summary outlines the features of periodically time-varying systems which are of major importance from an applications viewpoint. More extensive algebraic treatments will be found in the books by McLachlan [1] and Magnus and Winkler [2] both of which deal exclusively with second order systems. Arscott [3] has also provided a detailed exposition of second order periodic differential equations including the doubly-periodic Lamé equation, a topic given brief consideration also by Magnus and Winkler. A comprehensive matrix treatment of variable coefficient equations for general order systems has been given by D'Angelo [4]. Chapter 7 of that book, in particular is devoted to periodically varying equations. The recently translated Russian work by Yakubovich and Starzhinskii [5] also deals extensively with systems of higher orders containing periodically variable parameters.

The mode concept outlined above was developed by Keenan [6] and forms a convenient basis for the interpretation of the behaviour of periodic systems. This feature is considered further in Chap. 4. In particular *mode diagrams*, which summarise system modes over ranges of parameter values are presented and techniques for their generation are described. Also developed in Chap. 4 is the *stability diagram* which is a map of system stability over suitable ranges of parameters. This artifice was introduced by Ince in 1925 with regard to the Mathieu equation. Together, mode and stability diagrams provide a powerful summary of the behaviour of the periodically varying system. Whilst the latter permits a ready assessment of stability, the mode diagram allows predictions to be made concerning types of solution and thus, to a first order, harmonic content. This is important in circuit applications of periodic equations.

References for Chapter 2

1. McLachlan, N. W.: Theory and application of Mathieu functions. Clarendon: Oxford U. P. 1947. Reprinted by Dover, New York 1964
2. Magnus, W.; Winkler, S.: Hill's equation. New York: Wiley 1966
3. Arscott, F. M.: Periodic differential equations. Oxford: Pergamon 1964
4. D'Angelo, H.: Linear time-varying systems: Analysis and synthesis. Boston: Allyn & Bacon 1970
5. Yakubovich, V. A.; Starzhinskii, V. M.: Linear differential equations with periodic coefficients, Vol. 1 and 2. New York: Wiley 1975. (Translation from Russian)
6. Keenan, R. K.: An investigation into some problems in periodically parametric systems. PhD Thesis, Monash University Melbourne 1966

Problems

2.1 Verify Eq. (2.26a). Why is it applicable to both lossless and lossy second order periodic systems?

2.2 Determine the discrete transition matrix and characteristic exponents for the system described by the constant coefficient equation

$$\ddot{x} - a^2 x = 0.$$

Choose a time period of unity.

2.3 The discrete transition matrix for a second order system described by a Hill equation with the impulsive coefficient shown in Fig. 3.7 (with $\tau \to 0$) is

$$\phi(\pi, 0) = \begin{bmatrix} \cos d\pi - \frac{1}{d}\sin d\pi & \frac{1}{d}\sin d\pi \\ -(d\sin d\pi + \cos d\pi) & \cos d\pi \end{bmatrix}$$

where d^2 is a constant component in the coefficient. Determine the modes of the system for $d = 0.5, 1.0, 2.5, 5.0$.

2.4 Although the Mathieu equation is analytically intractable, its first order counterpart can be handled easily by well-known integration techniques. Show that the solution to

$$\dot{x} + (a - 2q\cos 2t)x = 0$$

is of the form

$$x(t) = A e^{-at} e^{2q\sin t}.$$

If a is regarded as its characteristic exponent what modes does a system described by this equation have?

2.5 Suppose $\psi(t)$ in Eq. (2.5) has a period 2π. Show that the change of time variable $\xi = t/2$ can be used to convert it to canonical form with period π. How are the values of a and q affected by such a transformation?

2.6 Verify Eq. (2.17b) by induction.

2.7 Show that complex eigenvalues for second order periodic systems must exist in conjugate pairs.

2.8 Demonstrate that the characteristic exponents for C modes in a second order, lossless periodic system must be purely imaginary. For a lossy system show that they take on a real part which is the same for both.

2.9 Using Eq. (2.27b) demonstrate that if complex (as against purely real) characteristic exponents exist for a Hill equation then their imaginary parts must be of the form $m\pi/\theta$, where m is an integer.

Chapter 3

Solutions to Periodic Differential Equations

Closed form solutions of periodic differential equations are, in general, difficult to find. Whilst series expansions can be determined in some cases [1] they are generally of little value in most applications, especially when many solutions are required. There are, however, a few specific types of periodic equation that can be solved analytically and for which the solutions appear in an easily used form. So useful and simple in fact are these special cases that they form the essence of modelling techniques which can be used to generate very good approximations to the solutions of intractable periodic differential equations. These modelling methods are the subject of Chap. 5. In this chapter equations which are tractable are treated in depth using the matrix approach laid down in Chap. 2. Methods for handling homogeneous equations are dealt with first whilst particular integrals are considered in a later section.

3.1 Solutions Over One Period of the Coefficient

Equations (2.17) describe a particularly important property of periodically time-varying equations, viz. that once the solution to a homogeneous form is known over one period (presumably $0 \leq t < \theta$) then it can be found easily for all time, simply by matrix multiplication with the discrete transition matrix $C = \phi(\theta, 0)$. Therefore the confronting task in solving such an equation is to determine both this matrix and the solution over one period. (Keenan [2] refers to the latter as a *first cycle solution.*) Fortunately both can be found with the same effort.

The key to determining the discrete transition matrix and the first cycle solution lies in the fact that the transition matrix over an interval is simply the product of transition matrices over a complete set of sub-intervals. For example

$$\begin{aligned} \phi(\theta, 0) &= \phi(\theta, t_n)\phi(t_n, t_{n-1}) \cdots \phi(t_1, 0), \\ &\qquad \theta > t_n > \cdots t_1 > 0. \end{aligned} \tag{3.1}$$

Thus if the $\phi(t_j, t_i)$ are known then $\phi(\theta, 0)$ and any transition matrix over a smaller interval can be found, the latter leading to the first cycle solution given, from Eq. (2.20), as $x(t) = \phi(t, 0)x(0)$, $0 \leq t < \theta$. To solve a periodic differential equation it is necessary therefore to identify the t_i, t_j in one period of the time variation such that over the intervals (t_i, t_j) the $\phi(t_j, t_i)$ can be found easily. This procedure will be illustrated in the following sections which treat explicit forms of Hill equations.

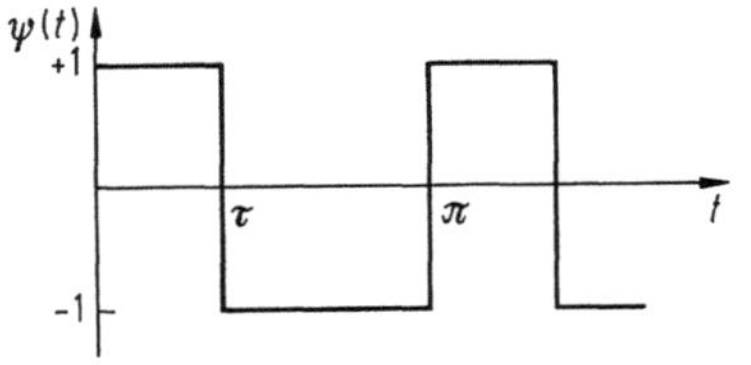

Fig. 3.1. Unit rectangular waveform coefficient in the Meissner equation

3.2 The Meissner Equation

Consider the second order equation

$$\ddot{x} + (a - 2q\psi(t))x = 0, \quad \psi(t) = \psi(t + \pi) \tag{3.2}$$

in which $\psi(t)$ is unit rectangular waveform depicted in Fig. 3.1. It is evident that over one period this equation can be viewed as the pair of constant coefficient equations

$$\ddot{x} + (a - 2q)x = 0 \quad 0 \leq t < \tau \tag{3.3a}$$

and

$$\ddot{x} + (a + 2q)x = 0 \quad \tau \leq t < \pi \tag{3.3b}$$

for which solutions are easily determined in terms of trigonometric functions. As a result, state transition matrices and thus complete and exact solutions can be found. To simplify notation let Eqs. (3.3) be rewritten as

$$\ddot{x} + c^2 x = 0$$

for which linearly independent basis solutions $x_i(t)$ are $\cos ct$ and $\sin ct$. The second basis function is generally chosen as $\frac{1}{c}\sin ct$ since this leads to slightly simpler notation. The corresponding Wronskian matrix (Eq. (2.9)) is

$$W(t) = \begin{bmatrix} \cos ct & \frac{1}{c}\sin ct \\ -c\sin ct & \cos ct \end{bmatrix}.$$

The state transition matrix for this equation over the time interval (t_1, t_2) is, according to Eq. (2.12)

$$\phi(t_2, t_1) = W(t_2)W^{-1}(t_1)$$

with

$$W^{-1}(t_1) = W^*(t_1)/\det W(t_1) = W^*(t_1)/W_0$$

in which $W^*(t_1)$ is the adjoint matrix. Thus

$$W^{-1}(t_1) = \begin{bmatrix} \cos ct_1 & -\frac{1}{c}\sin ct_1 \\ c\sin ct_1 & \cos ct_1 \end{bmatrix}$$

since $W_0 = 1$. Therefore

$$\phi(t_2, t_1) = \begin{bmatrix} \cos c(t_2 - t_1) & \frac{1}{c}\sin c(t_2 - t_1) \\ -c\sin c(t_2 - t_1) & \cos c(t_2 - t_1) \end{bmatrix}. \tag{3.4}$$

Applying this result to Eqs. (3.3) gives as the state transition matrix for the Meissner equation

$$\phi(\pi, 0) = \begin{bmatrix} \cos d(\pi - \tau) & \frac{1}{d}\sin d(\pi - \tau) \\ -d\sin d(\pi - \tau) & \cos d(\pi - \tau) \end{bmatrix} \begin{bmatrix} \cos c\tau & \frac{1}{c}\sin c\tau \\ -c\sin c\tau & \cos c\tau \end{bmatrix} \tag{3.5a}$$

with

$$d = \sqrt{a + 2q} \quad \text{and} \quad c = \sqrt{a - 2q}. \tag{3.5b}$$

Here the significant time intervals have been seen to be $(0, \tau)$ and (τ, π).

In a similar manner the transition matrix over an interval smaller than π can be determined.

3.3 Solution at Any Time for a Second Order Periodic Equation

It is instructive to continue the above example. However the analysis to follow holds for any second-order (Hill) equation and thus it is useful to separate it. Equations (3.6) and (3.7) to follow apply also to higher order equations.

At a time $t = m\pi + \chi$, $\chi < \pi$, m-integral, the solution to a homogeneous periodic differential equation can be seen from Eqs. (2.16) and (2.20) to be

$$\boldsymbol{x}(t) = \phi(\chi, 0)\phi(\pi, 0)^m \boldsymbol{x}(0) \tag{3.6}$$

where $\phi(\chi, 0)$ is the transition matrix over an interval smaller than π. It is convenient to diagonalise $\phi(\pi, 0)$ according to

$$\phi(\pi, 0) = \psi\Lambda\psi^{-1}$$

where Λ is a diagonal matrix of eigenvalues of $\phi(\pi, 0)$ and ψ is a matrix of eigenvectors. (For the purpose of this section the eigenvalues will be considered as distinct). Equation (3.6) therefore becomes

$$\boldsymbol{x}(t) = \phi(\chi, 0)\psi\Lambda^m\psi^{-1}\boldsymbol{x}(0). \tag{3.7}$$

The elements of ψ can be found from the equations

$$\{\phi(\pi, 0) - \lambda_i I\}\psi_i = 0$$

where I is an identity matrix and ψ_i is the eigenvector associated with the eigenvalue λ_i. These equations are a homogeneous set, different for each λ_i, and a unique solution does not exist. Therefore one unknown can be assigned arbitrarily. Putting $\psi_{1i} = 1$, $i = 1, 2$, gives for a second-order Hill equation

$$\psi_{2i} = \frac{\lambda_i - \phi_{11}}{\phi_{12}}, \tag{3.8}$$

thereby defining the elements of the matrix ψ by putting $i = 1$ and 2 respectively.

When Eq. (3.8) is substituted into Eq. (3.7) it will be seen that the solution to the Hill equation, in algebraic form, is of the type

$$x(t) = \lambda_1^m \Xi_1(\chi) + \lambda_2^m \Xi_2(\chi) \tag{3.9}$$

which should be compared with Eqs. (2.21) and (2.25). $\dot{x}(t)$ will be similar in form.

Equation (3.9) is a convenient expression with which to evaluate $x(t)$. The functions $\Xi_1(\chi)$ and $\Xi_2(\chi)$ are periodic with period π and thus need only be found over the first period $0 \leq t < \pi$. Thereafter the solution in subsequent periods is found simply by multiplying repetitively by the eigenvalues as indicated. It is a relatively straightforward matter to show

$$\Xi_i(\chi) = \frac{k_i}{\lambda_2 - \lambda_1}\{a_{11} + a_{12}(\lambda_i - \phi_{11})/\phi_{12}\} \quad i = 1, 2 \tag{3.10}$$

in which ϕ_{11} and ϕ_{12} are elements of $\phi(\pi, 0)$, a_{11} and a_{12} are elements of $\phi(\chi, 0)$ and

$$k_1 = (\lambda_2 - \phi_{11})x(0) - \phi_{12}\dot{x}(0),$$

$$k_2 = (\phi_{11} - \lambda_1)x(0) + \phi_{12}\dot{x}(0).$$

3.4 Evaluation of $\phi(\pi, 0)^m$, m Integral

In the preceding section, the term $\phi(\pi, 0)^m$ appearing in Eq. (3.6) has been replaced by the operation of raising the eigenvalues of $\phi(\pi, 0)$ to the mth power. In applications demanding a complete continuous record of solution up until a particular time this is clearly a desirable technique. If, however, the solution is needed only *after* a particular number of periods, m, have elapsed a technique for evaluating $\phi(\pi, 0)^m$ directly, developed by Pipes [3], is particularly attractive. This depends upon an application of Sylvester's theorem [4] in the following manner. The polynomial matrix equation $\phi(\pi, 0)^m$ can be expressed as

$$\phi(\pi, 0)^m = \lambda_1^m H_0(\lambda_1) + \lambda_2^m H_0(\lambda_2) \tag{3.11}$$

in which

$$H_0(\lambda_i) = F(\lambda_i) \Big/ \left(\frac{\mathrm{d}\Delta}{\mathrm{d}\lambda}\right)_{\lambda=\lambda_i}$$

where $F(\lambda_i)$ is the adjoint of the characteristic matrix of $\phi(\pi, 0)$ and $\Delta(\lambda)$ is the characteristic polynomial.

The characteristic matrix is

$$[\lambda I - \phi(\pi, 0)] = \begin{bmatrix} \lambda - \phi_{11} & -\phi_{12} \\ -\phi_{21} & \lambda - \phi_{22} \end{bmatrix}$$

with adjoint

$$\begin{bmatrix} \lambda - \phi_{22} & \phi_{12} \\ \phi_{21} & \lambda - \phi_{11} \end{bmatrix}$$

Table 3.1. Evaluation of $\phi(\pi, 0)^m$

Distinct eigenvalues

(i) $\phi_{11} + \phi_{22} \neq \pm 2$

$$\phi(\pi, 0)^m = \frac{1}{s_1}\begin{bmatrix} s_{m+1} - \phi_{22}s_m & \phi_{12}s_m \\ \phi_{21}s_m & s_{m+1} - \phi_{11}s_m \end{bmatrix}$$

in which $s_m = \sinh(mb)$

(ii) $\phi_{11} = \phi_{22} \neq \pm 1$. Therefore $\cosh(b) = \dfrac{T}{2} = \phi_{11}$. (This case is encountered over a constant coefficient, amongst other special cases of more general coefficients.)

$$\phi(\pi, 0)^m = \begin{bmatrix} c_m & Z_0 s_m \\ \dfrac{1}{Z_0}s_m & c_m \end{bmatrix}$$

in which $s_m = \sinh(mb)$, $c_m = \cosh(mb)$ and $Z_0 = [\phi_{12}/\phi_{21}]^{1/2}$

Degenerate eigenvalues

(i) $\phi_{11} + \phi_{12} = +2$, $\lambda_1 = \lambda_2 = 1$

$$\phi(\pi, 0)^m = \begin{bmatrix} m(\phi_{11} - 1) + 1 & m\phi_{12} \\ m\phi_{21} & m(\phi_{22} - 1) + 1 \end{bmatrix}$$

(ii) $\phi_{11} + \phi_{12} = -2$, $\lambda_1 = \lambda_2 = -1$

$$\phi(\pi, 0)^m = mE_{m-1}\begin{bmatrix} (\phi_{11} + 1) & \phi_{12} \\ \phi_{21} & (\phi_{22} + 1) \end{bmatrix} + E_{m+1}I$$

where $E_m = \exp(jm\pi)$

whereas the characteristic polynomial is, from Eq. (2.26b),

$$\Delta = \det[\phi(\pi, 0) - \lambda I] = \lambda^2 - \lambda(\phi_{11} + \phi_{22}) + 1.$$

Thus

$$\frac{d\Delta}{d\lambda} = 2\lambda - \text{trace}\{\phi(\pi, 0)\}. \tag{3.12}$$

The values of the λ_i are given as the roots of the characteristic equation and are thus

$$\lambda_{1,2} = \frac{T}{2} \pm \sqrt{\left(\frac{T}{2}\right)^2 - 1} \tag{3.13}$$

in which $T = \text{trace}\{\phi(\pi, 0)\}$.

If

$$\frac{T}{2} \overset{\text{def}}{=} \cosh(b)$$

then

$$\lambda_{1,2} = \cosh(b) \pm \sinh(b)$$

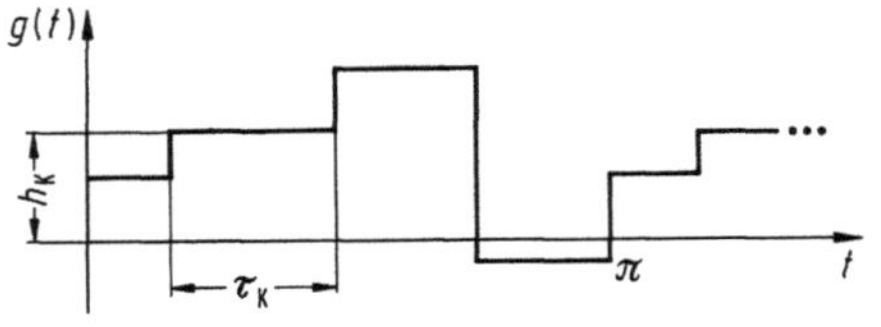

Fig. 3.2. Periodic staircase coefficient

so that

$$\lambda_1 = e^b \quad \text{and} \quad \lambda_2 = e^{-b}. \tag{3.14}$$

$$\text{For} \quad T = +2, \quad b = 0 \quad \text{and} \quad \lambda_1 = \lambda_2 = 1,$$

and

$$\text{for} \quad T = -2, \quad b = 0 \quad \text{and} \quad \lambda_1 = \lambda_2 = -1,$$

as seen from Eq. (3.13).

Using these relationships, the results of Table 3.1 (taken from Pipes [3]) can be derived.

The forms of Eq. (3.11) and the subsequent supporting expressions should be compared to the development in the preceding section.

3.5 The Hill Equation with a Staircase Coefficient

In the modelling procedures to be developed in Chap. 5 Hill equations of the form

$$\ddot{x} + g(t)x = 0$$

in which $g(t)$ is a staricase or piecewise-constant coefficient as illustrated in Fig. 3.2, are of central importance.

Equation (3.4) shows that over one step of the coefficient the state transition matrix is

$$\phi(t + \tau_k, t) = \phi(\tau_k, 0) = \begin{bmatrix} \cos c\tau_k & \frac{1}{c}\sin c\tau_k \\ -c\sin c\tau_k & \cos c\tau_k \end{bmatrix} \tag{3.15}$$

where $c = \sqrt{h_k}$. State transition matrices over all steps, or even parts of steps, in one period of the coefficient can be determined similarly, permitting a complete analytical solution to the Hill equation in question.

3.6 The Hill Equation with a Sawtooth Waveform Coefficient

Another tractable Hill equation is that in which the coefficient $\psi(t)$ is a sawtooth function. Pipes [3] appears to have been the first to demonstrate the solution to this type of equation although the basis functions and subsequent theory outlined below are slightly different to those employed in his work.

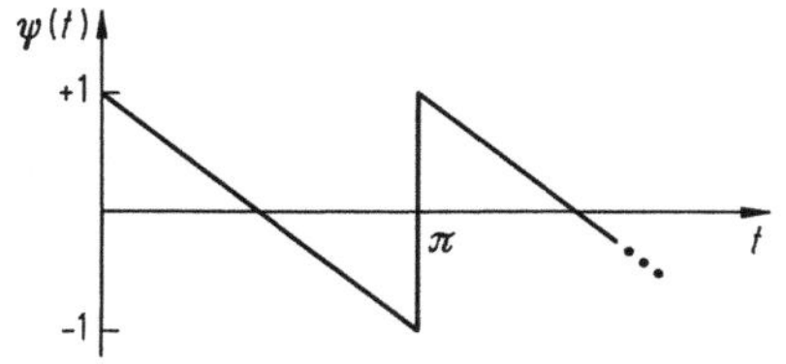

Fig. 3.3. Negative sloping sawtooth waveform coefficient

Consider the negative-slope sawtooth coefficient $\psi(t)$ illustrated in Fig. 3.3. Over one period a Hill equation with this periodic coefficient is

$$\ddot{x} + \{a - 2q(-2t/\pi + 1)\}x = 0, \quad 0 \leq t < \pi. \tag{3.16}$$

It is convenient to put $\alpha = a - 2q$ and $\beta = +4q/\pi$, so that the change of variable

$$z = \alpha + \beta t$$

changes Eq. (3.16) to Stoke's equation

$$\frac{d^2x}{dz^2} + \frac{z}{\beta^2}x = 0$$

which has the solution [3]

$$x(z) = \sqrt{z}\{AJ_{1/3}(\sigma) + BJ_{-1/3}(\sigma)\}$$

in which $J_{\pm 1/3}(\sigma)$ is the Bessel function of the first kind of order $\pm 1/3$ and $\sigma = 2z^{3/2}(3\beta)^{-1}$. A set of basis solutions therefore over one period of the Hill equation with a triangular coefficient is

$$x_1(t) = \sqrt{z}J_{1/3}(\sigma) \tag{3.17a}$$

$$x_2(t) = \sqrt{z}J_{-1/3}(\sigma). \tag{3.17b}$$

To construct the corresponding Wronskian matrix $\dot{x}_1(t)$ and $\dot{x}_2(t)$ are now required. These can be found in the following manner in which $x_2(t)$ has been chosen as an illustration.

$$\frac{dx_2(t)}{dz} = J_{-1/3}(\sigma)/2\sqrt{z} + \sqrt{z}\frac{d\sigma}{dz}\frac{d}{d\sigma}J_{-1/3}(\sigma).$$

A recurrence relation for Bessel functions[1] shows

$$\frac{d}{d\sigma}J_{-1/3}(\sigma) = -\frac{1}{3\sigma}J_{-1/3}(\sigma) - J_{2/3}(\sigma)$$

so that

$$\frac{dx_2(t)}{dz} = -\frac{z}{\beta}J_{2/3}(\sigma).$$

[1] McLachlan [5], p. 191, Eq. (25).

Therefore

$$\frac{dx_2(t)}{dt} = -\frac{z}{\beta}J_{2/3}\sigma\frac{dz}{dt} = -zJ_{2/3}(\sigma). \tag{3.18a}$$

In a similar manner it can be shown that

$$\frac{dx_1(t)}{dt} = zJ_{-2/3}(\sigma) \tag{3.18b}$$

so that the Wronskian matrix is

$$W(t) = \begin{bmatrix} \sqrt{z}J_{1/3}(\sigma) & \sqrt{z}J_{-1/3}(\sigma) \\ zJ_{-2/3}(\sigma) & -zJ_{2/3}(\sigma) \end{bmatrix}. \tag{3.19}$$

Furthermore

$$W^*(t) = \begin{bmatrix} -zJ_{2/3}(\sigma) & -\sqrt{z}J_{-1/3}(\sigma) \\ -zJ_{-2/3}(\sigma) & \sqrt{z}J_{1/3}(\sigma) \end{bmatrix}. \tag{3.20}$$

The Wronskian for the chosen basis solutions is

$$W_0 = -z\sqrt{z}\{J_{1/3}(\sigma)J_{2/3}(\sigma) + J_{-1/3}(\sigma)J_{-2/3}(\sigma)\}.$$

This can be shown to be a constant in the following way.

$$J_{\pm 2/3}(\sigma) = \mp \frac{1}{3\sigma}J_{\mp 1/3}(\sigma) \mp J'_{1/3}(\sigma)\ ^{2}$$

so that the Wronskian becomes

$$\begin{aligned} W_0 &= z\sqrt{z}\,\{J_{1/3}(\sigma)J'_{-1/3}(\sigma) - J_{-1/3}(\sigma)J'_{1/3}(\sigma)\} \\ &= -3\sqrt{3}\beta/2\pi.\ ^{3} && (3.21\text{a}) \\ &= -6\sqrt{3}q/\pi^2. && (3.21\text{b}) \end{aligned}$$

By an asymptotic expansion Pipes [3] derives the Wronkian for this example as $3\beta/\pi$ where his choice of basis function is $\sqrt{z}J_{1/3}(\sigma)$ and $\sqrt{z}Y_{1/3}(\sigma)$, Y being the Bessel function of the second kind.

With the above information the transition matrix over an interval within the range $(0, \pi)$ can be derived. In particular, the discrete transition matrix will be given as

$$\begin{aligned} \phi(\pi, 0) &= -2\pi W(\pi)W^*(0)/3\sqrt{3}\beta \\ &= \frac{-\pi^2}{6\sqrt{3}q}\begin{bmatrix} dJ_{1/3}(kd^3) & dJ_{-1/3}(kd^3) \\ d^2J_{-2/3}(kd^3) & -d^2J_{2/3}(kd^3) \end{bmatrix}\begin{bmatrix} -c^2J_{2/3}(kc^3) & -cJ_{-1/3}(kc^3) \\ -c^2J_{-2/3}(kc^3) & cJ_{1/3}(kc^3) \end{bmatrix} \end{aligned}$$

[2] See McLachlan [5], p. 191. Eqs. (25) and (26).
[3] See McLachlan [5], p. 192, Eq. (35).

where $d^2 = a + 2q$, $c^2 = a - 2q$, $k = \dfrac{\pi}{6q}$.

Thus

$$\phi(\pi, 0) = \frac{\pi^2}{6\sqrt{3}q}\begin{bmatrix} dc^2\{J_{1/3}(\sigma_d)J_{2/3}(\sigma_c) + J_{-1/3}(\sigma_d)J_{-2/3}(\sigma_c)\} & cd\{J_{1/3}(\sigma_d)J_{-1/3}(\sigma_c) - J_{-1/3}(\sigma_d)J_{1/3}(\sigma_c)\} \\ c^2d^2\{J_{-2/3}(\sigma_d)J_{2/3}(\sigma_c) - J_{2/3}(\sigma_d)J_{-2/3}(\sigma_c)\} & cd^2\{J_{-2/3}(\sigma_d)J_{-1/3}(\sigma_c) + J_{2/3}(\sigma_d)J_{1/3}(\sigma_c)\} \end{bmatrix} \tag{3.22a}$$

where $\sigma_d = kd^3$ and $\sigma_c = kc^3$.
For later reference, write this as

$$\phi_-(\pi, 0) = \frac{\pi^2}{6\sqrt{3}q} M_-(k). \tag{3.22b}$$

In practice the parameters z and σ may not always be positive. Consideration of their definitions will show that both can be negative and, in addition, z can be zero. It is necessary therefore that the elements of the Wronskian matrix be suitably modified in those circumstances. Using the recurrence relations and identities found in McLachlan [5] the following special cases can be derived.

3.6.1 The Wronskian Matrix with z Negative

$$W_{11}(-z) = -\sqrt{z}I_{1/3}(\sigma)$$
$$W_{12}(-z) = \sqrt{z}I_{-1/3}(\sigma)$$
$$W_{21}(-z) = zI_{-2/3}(\sigma)$$
$$W_{22}(-z) = -zI_{2/3}(\sigma)$$

in which $I_{\pm 1/3}(\ )$ and $I_{\pm 2/3}(\ )$ are modified Bessel functions of the first kind.

3.6.2 The Wronskian Matrix with z Zero

$$W_{11}(0) = 0$$
$$W_{12}(0) = (3\beta)^{1/3}/\Gamma(2/3)$$
$$W_{21}(0) = (3\beta)^{2/3}/\Gamma(1/3)$$
$$W_{22}(0) = 0$$

in which $\Gamma(2/3)$ and $\Gamma(1/3)$ are gamma functions. It may be noted that

$$\Gamma(1/3) = 2.6789385347$$

and

$$\Gamma(2/3) = 1.3541179394$$ [4]

---

[4] Abramowitz and Stegun [6], p. 255

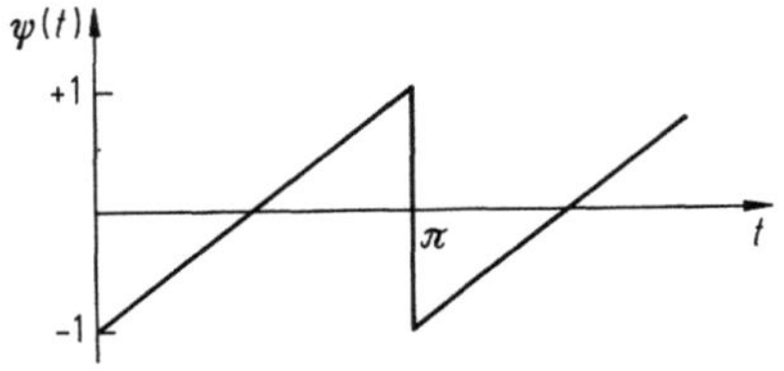

Fig. 3.4. Positive sloping sawtooth waveform coefficient

3.6.3 The Case of β Negative

For this situation it is better to look at the elements of the transition matrix rather than those of the Wronskian matrix. If σ_1 and σ_2 are the values of σ at the end and beginning respectively of a time interval within one period of a ramp coefficient then the following Bessel function products occur, as seen in Eq. (3.22),

$$J_{1/3}(\sigma_1)J_{2/3}(\sigma_2)$$
$$J_{-1/3}(\sigma_1)J_{-2/3}(\sigma_2)$$
$$J_{2/3}(\sigma_1)J_{1/3}(\sigma_2)$$
$$J_{-2/3}(\sigma_1)J_{-1/3}(\sigma_2)$$
$$J_{1/3}(\sigma_1)J_{-1/3}(\sigma_2)$$
$$J_{-1/3}(\sigma_1)J_{1/3}(\sigma_2)$$
$$J_{2/3}(\sigma_1)J_{-2/3}(\sigma_2)$$
$$J_{-2/3}(\sigma_1)J_{2/3}(\sigma_2).$$

It can be shown that when k is negative the only effect is that the first four products change in sign.

When z is negative as well these results must also incorporate a change in sign and the use of modified Bessel functions.

3.7 The Hill Equation with a Positive Slope, Sawtooth Waveform Coefficient

For the coefficient $\psi(t)$ illustrated in Fig. 3.4. the Hill equation, over one period appears as

$$\ddot{x} + \{a - 2q(2t/\pi - 1)\}x = 0 \quad 0 \leq t < \pi$$

so that putting $\alpha = a + 2q$ and $\beta = -4q/\pi$ with $z = \alpha + \beta t$ allows Eqs. (3.19) and (3.20) to be used again to generate transition matrices. In particular, the discrete transition matrix will again be given by Eq. (3.22) but with the modification of Sec. 3.6.3 to account for β being negative along with a change of sign for the Wronskian. Thus

$$\phi(\pi, 0) = \frac{-\pi^2}{6\sqrt{3q}}\begin{bmatrix} -dc^2\{J_{1/3}(\sigma_d)J_{2/3}(\sigma_c) + J_{-1/3}(\sigma_d)J_{-2/3}(\sigma_c)\} & cd\{J_{1/3}(\sigma_d)J_{-1/3}(\sigma_c) - J_{-1/3}(\sigma_d)J_{1/3}(\sigma_c)\} \\ c^2d^2\{_{-2/3}(\sigma_d)J_{2/3}(\sigma_c) - J_{2/3}(\sigma_d)J_{-2/3}(\sigma_c)\} & -cd^2\{J_{-2/3}(\sigma_d)J_{-1/3}(\sigma_c) + J_{2/3}(\sigma_d)J_{1/3}(\sigma_c)\} \end{bmatrix} \quad (3.23a)$$

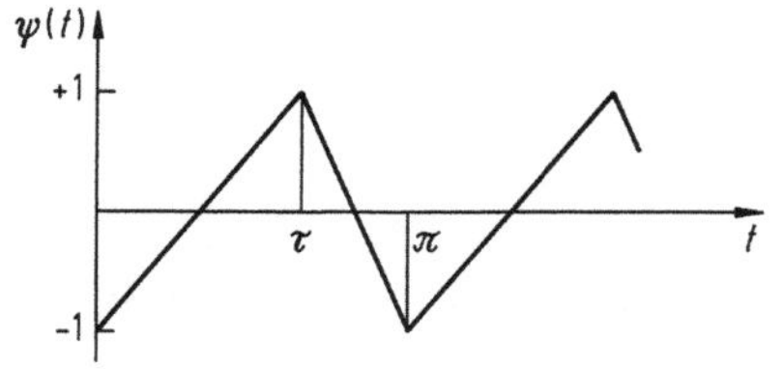

Fig. 3.5. Triangular waveform coefficient

i.e.

$$\phi_+(\pi, 0) = \frac{-n^2}{6\sqrt{3q}} M_+(k) \tag{3.23b}$$

with

$$\sigma_d = kd^3, d^2 = a - 2q$$

and

$$\sigma_c = kc^3, c^2 = a + 2q$$

$$\text{with } k = \frac{\pi}{6q}$$

3.8 The Hill Equation with a Triangular Coefficient

When the Hill equation has the triangular coefficient shown in Fig. 3.5 the results of Eqs. (3.22) and (3.23) can be used together to produce the discrete transition matrix or a transition matrix over an interval within $(0, \pi)$. The discrete transition matrix, in particular, will be given by

$$\phi(\pi, 0) = \phi_-(\pi, \tau)\,\phi_+(\tau, 0) = \phi_-(\pi - \tau, 0)\phi_+(\tau, 0)$$

where the subscripts $-$ and $+$ refer to the negative and positive gradient portions of the triangle respectively; $\phi_-(\pi, \tau)$ can be deduced from Eq. (3.22) whilst $\phi_+(\tau, 0)$ can be derived from Eq. (3.23).

The triangular coefficient is described by

$$\psi(t) = \frac{2}{\tau}t - 1 \qquad 0 < t \leq \tau$$

$$\frac{-2}{\pi - \tau}t' + 1 \quad 0 < t' \leq \pi - \tau, t' = t - \tau.$$

When inserted into the Hill equation these give rise to the transition matrices:

$$\phi_-(\pi - \tau, 0) = \frac{+\pi(\pi - \tau)}{6\sqrt{3q}} M_-(k), \quad k = (\pi - \tau)/6q$$

and

$$\phi_+(\tau, 0) = \frac{-\pi\tau}{6\sqrt{3q}} M_+(k), \quad k = \tau/6q.$$

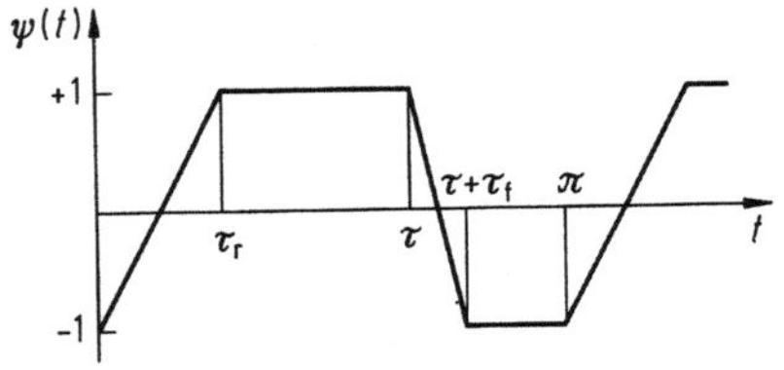

Fig. 3.6. Periodic trapezoidal waveform coefficient

3.9 The Hill Equation with a Trapezoidal Coefficient

One of the most useful forms of the Hill equation, particularly from the point of view of constructing a model for the Mathieu equation (see Chap. 6), is that in which the coefficient is a periodic trapezoid as depicted in Fig. 3.6. The discrete transition matrix in this case is

$$\phi(\pi, 0) = \phi_L(\pi, \tau + \tau_f)\phi_-(\tau + \tau_f, \tau)\phi_U(\tau, \tau_r)\phi_+(\tau_r, 0)$$

in which $\phi_+(\, , \,)$ and $\phi_-(\, , \,)$ have the same meanings as in the preceding section whereas $\phi_U(\, , \,)$ and $\phi_L(\, , \,)$ are transition matrices over the upper and lower constant parts of the trapezoid respectively.

It is convenient to rewrite the discrete transition matrix as

$$\begin{aligned}\phi(\pi, 0) &= \phi_L(\pi - \tau - \tau_f, 0)\phi_-(\tau_f, 0)\phi_U(\tau - \tau_r, 0)\phi_+(\tau_r, 0)\\ &= -\phi_L(\pi - \tau - \tau_f, 0)\frac{\pi\tau_f}{6\sqrt{3q}}M_-\left(\frac{\tau_f}{6q}\right)\\ &\quad \phi_U(\tau - \tau_r, 0)\frac{\pi\tau_r}{6\sqrt{3q}}M_+\left(\frac{\tau_r}{6q}\right)\end{aligned}$$

where M_- and M_+ are defined in Eqs. (3.22) and (3.23) and in which $\phi_L(\pi - \tau - \tau_f, 0)$ and $\phi_U(\tau - \tau_r, 0)$ are given by Eq. (3.15) as

$$\phi_L(\pi - \tau - \tau_f, 0) = \begin{bmatrix} \cos d(\pi - \tau - \tau_f) & \frac{1}{d}\sin d(\pi - \tau - \tau_f) \\ -d\sin d(\pi - \tau - \tau_f) & \cos d(\pi - \tau - \tau_f) \end{bmatrix}$$

with

$$d = \sqrt{a + 2q}$$

and

$$\phi_U(\tau_r, 0) = \begin{bmatrix} \cos c\tau_r & \frac{1}{c}\sin c\tau_r \\ -c\sin c\tau_r & \cos c\tau_r \end{bmatrix}$$

with $c = \sqrt{a - 2q}$.

3.10 Bessel Function Generation

The foregoing sections demonstrate that Hill equations with coefficients composed of or containing ramps demand the availability of means for evaluating the

Bessel functions $J_{\pm 1/3}$, $J_{\pm 2/3}$, $I_{\pm 2/3}$. Whilst these could be read from tables, it is desirable in general to evaluate them as the need arises to avoid errors introduced by interpolating within a table and to preclude having to have available (in computer memory for example) extensive tabulations. Direct generation using the series expansion definitions however is undesirable for two reasons. First fractional exponentiation is required in the case of a Bessel function of fractional order. This is a time-consuming operation, computationally. Secondly it is difficult, with the series definition, to obtain an estimate of the error made by truncating the series at a particular term. Both of these difficulties are circumvented fortunately by the use of Chebyshev series techniques for functionally approximating the required Bessel functions. This method is outlined in the Appendix.

3.11 The Hill Equation with a Repetitive Exponential Coefficient

If the periodic coefficient in a Hill equation has segments which are exponentially time-varying, then over the corresponding time intervals the equation will appear as

$$\ddot{x} + (\alpha + \beta e^{-\sigma t})x = 0. \tag{3.24}$$

As an illustration, Eq. (3.24) would arise in the Meissner equation, Eqs. (3.2), (3.3), if the rectangular waveform were low-pass filtered.

With the change of variable

$$z = 2\sqrt{\beta}e^{-\sigma t/2}/\sigma$$

Eq. (3.24) becomes

$$\frac{d^2x}{dz^2} + \frac{1}{z}\frac{dx}{dz} + (1 - \nu^2/z^2)x = 0$$

which is Bessel's equation with $\nu^2 = -4\alpha/\sigma^2$. This has the fundamental solutions

$$x_1(t) = J_\nu(z) \quad \text{and} \quad x_2(t) = Y_\nu(z).$$

To determine the Wronskian and state transition matrices the time derivatives of these are necessary. Thus

$$\frac{dx_1(t)}{dt} = \frac{d}{dz}J_\nu(z)\frac{dz}{dt}$$
$$= -\tfrac{\sigma}{2}zJ'_\nu(z)$$

and

$$\frac{dx_2(t)}{dt} = -\tfrac{\sigma}{2}zY'_\nu(z)$$

so that the Wronskian matrix is

$$W(t) = \begin{bmatrix} J_\nu(z) & Y_\nu(z) \\ -\tfrac{\sigma}{2}zJ'_\nu & -\tfrac{\sigma}{2}zY'_\nu(z) \end{bmatrix}$$

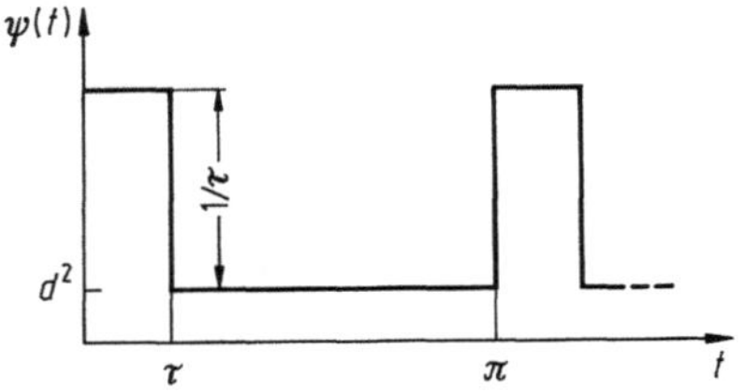

Fig. 3.7. Pulse waveform coefficient. In the limit as $\tau \to 0$ this becomes a periodic sequence of impulses superimposed on a constant of value d^2. Note that q has no sensible meaning here, and that $a = d^2$

giving as the Wronskian (using formula 115 of McLachlan [5], p. 197)

$$W_0 = \tfrac{\sigma}{2} z\{Y_\nu(z)J'_\nu(z) - J_\nu(z)Y'_\nu(z)\} = -\tfrac{\sigma}{\pi}$$

3.12 The Hill Equation with a Coefficient in the Form of a Repetitive Sequence of Impulses

Consider a coefficient in Hill's equation of the form shown in Fig. 3.7. This will transform into a constant plus a periodic sequence of impulses in the limit as $\tau \to 0$.

Equation (3.4) shows the discrete transition matrix associated with this coefficient to be

$$\phi(\pi, 0) = \begin{bmatrix} \cos d(\pi - \tau) & \frac{1}{d}\sin d(\pi - \tau) \\ -d \sin d(\pi - \tau) & \cos d(\pi - \tau) \end{bmatrix} \begin{bmatrix} \cos c\tau & \frac{1}{c}\sin c\tau \\ -c \sin c\tau & \cos c\tau \end{bmatrix}$$

where $c = \sqrt{d^2 + \frac{1}{\tau}}$.

For τ small $c \simeq \tau^{-1/2}$ so that

$$\phi(\pi, 0) = \begin{bmatrix} \cos d(\pi - \tau) & \frac{1}{d}\sin d(\pi - \tau) \\ -d \sin d(\pi - \tau) & \cos d(\pi - \tau) \end{bmatrix} \begin{bmatrix} \cos \tau^{1/2} & \tau^{1/2} \sin \tau^{1/2} \\ -\tau^{-1/2} \sin \tau^{1/2} & \cos \tau^{1/2} \end{bmatrix}$$

which in the limit as $\tau \to 0$ becomes

$$\phi(\pi, 0) = \begin{bmatrix} \cos d\pi & \frac{1}{d}\sin d\pi \\ -d \sin d\pi & \cos d\pi \end{bmatrix} \begin{bmatrix} 1 & 0 \\ -1 & 1 \end{bmatrix} \tag{3.25}$$

i.e.

$$\phi(\pi, 0) = \begin{bmatrix} \cos d\pi - \frac{1}{d}\sin d\pi & \frac{1}{d}\sin d\pi \\ -(d \sin d\pi + \cos d\pi) & \cos d\pi \end{bmatrix} \tag{3.26}$$

Equation (3.26) is a particularly simple expression for the discrete transition matrix for the Hill equation with a repetitive impulse-like coefficient and can now be used to evaluate solutions and to check stability. Had the impluse in the period π occurred after the constant portion (d^2) of the coefficient, rather than before as illustrated in Fig. 3.7, the component matrices in Eq. (3.25) would have appeared in the reverse order giving as the state transition matrix

$$\phi(\pi, 0) = \begin{bmatrix} \cos d\pi & \frac{1}{d}\sin d\pi \\ -(\cos d\pi + d\sin d\pi) & \cos d\pi - \frac{1}{d}\sin d\pi \end{bmatrix} \tag{3.27}$$

which is particularly interesting since it shows that over the first time period $(0, \pi)$ the *solution*, for the initial conditions $x(0)$ and $\dot{x}(0)$ at $t = 0$, is simply the solution to the constant coefficient equation which exists in the absence of the impulses. On the other hand the *derivative* at $t = \pi$ is different to the derivative just prior to the impulse, i.e. at $t = \pi^-$, showing the derivative to be discontinuous. These facts are seen by observing that the elements in the first row of Eq. (3.27) are those in the first row of the constant-coefficient matrix in Eq. (3.4) whereas the elements of the second rows are different. It is the first row elements of a transition matrix that determine solutions whereas elements in succeeding rows determine derivatives.

It is important to note that Eq. (3.25) can be generalised to

$$\phi(\pi, 0) = \begin{bmatrix} x_1(\pi) & x_2(\pi) \\ \dot{x}_1(\pi) & \dot{x}_2(\pi) \end{bmatrix} \begin{bmatrix} 1 & 0 \\ -1 & 1 \end{bmatrix} \tag{3.28}$$

where $x_1(t)$ and $x_2(t)$ are a pair of basis solutions valid over the interval $(0, \pi)$ corresponding to some general form of periodic coefficient, other than a constant, added to the impulses over that interval.

3.13 Equations of Higher Order

Whilst the majority of periodic differential equations encountered in practice are of second order there are certain situations, notably in electrical networks, where equations of higher order are found. These can be treated in an analogous manner to their second-order counterparts and again, the availability of an analytical solution depends upon being able to find basis solutions for an equation over a complete set of sub-intervals in one period of the periodic coefficients. A particular example would be the vth order equation with rectangularly-varying coefficients—i.e. a vth order Meissner equation. Over intervals wherein the coefficients are constant a set of basis functions would be

$$\{e^{s_i t}\} \quad i = 1, \ldots, v$$

where the s_i are the v roots of the appropriate characteristic equation. Thus by differentiating $(v - 1)$ times the relevant Wronskian matrix can be established and the state transition matrices ultimately determined.

3.14 Response to a Sinusoidal Forcing Function

Although an analytical solution to a forced (inhomogeneous) Hill equation depends upon tractability of the homogeneous form, some general results of practical significance can be derived at this stage.

The forced response of a general order periodic system, described for example by Eq. (2.7), is seen from Eq. (2.13) to be of the form

$$\boldsymbol{x}_F(t) = \int_0^t \phi(t, \tau)\boldsymbol{f}(\tau)\mathrm{d}\tau \tag{3.29}$$

where $\boldsymbol{f}(t)$ is a vector describing the nature of the excitation. In the case of sinusoidal forcing this will be of the form

$$\boldsymbol{f}(t) = \boldsymbol{d}\exp(\mathrm{j}\omega_i t). \tag{3.30}$$

Let $\phi(t, \tau) = \phi(t, 0)\phi(0, \tau) = \phi(t, 0)\phi^{-1}(\tau, 0)$ so that Eq. (3.29) with Eq. (3.30) can be written

$$\boldsymbol{x}_F(t) = \phi(t, 0)\int_0^t \phi^{-1}(\tau, 0)\boldsymbol{d}\exp(\mathrm{j}\omega_i\tau)\mathrm{d}\tau. \tag{3.31}$$

From Eq. (2.19)

$$\phi(t, 0) = P(t)\exp(\Gamma t)$$

in which $P(t)$ is periodic, with the period of the time-varying parameters, θ, and Γ is a matrix whose eigenvalues are the characteristic exponents of the system. Similarly $\phi^{-1}(\tau, 0) = P(t)^{-1}\exp(-\Gamma\tau)$, since $P(t)$ is non-singular (see Eq. (2.19) and note that $\phi(t, 0)$ must be non-singular) allowing Eq. (3.31) to be recast as

$$\boldsymbol{x}_F(t) = P(t)\exp(\Gamma t)\int_0^t P^{-1}(\tau)\exp(-\Gamma\tau)\boldsymbol{d}\exp(\mathrm{j}\omega_i\tau)\mathrm{d}\tau. \tag{3.32}$$

Since $P(t)$ is periodic $P^{-1}(\tau)$ is also periodic and can therefore be expressed as a Fourier series:

$$P^{-1}(\tau) = \sum_{k=-\infty}^{\infty} P_k\exp(\mathrm{j}k\omega_p\tau)$$

in which $\omega_p = 2\pi/\theta$. Equation (3.32) therefore becomes

$$\boldsymbol{x}_F(t) = P(t)\exp(\Gamma t)\int_0^t \sum_{k=-\infty}^{\infty} P_k\exp[\mathrm{j}(\omega_i + k\omega_p)\tau]\exp(-\Gamma\tau)\boldsymbol{d}\mathrm{d}\tau. \tag{3.33}$$

By the use of Sylvester's theorem (see Pipes [4])

$$\exp(-\Gamma\tau) = \sum_{l=1}^{v}\exp(-\mu_l\tau)\boldsymbol{Z}_l$$

in which the μ_l are the (distinct) eigenvalues of Γ, and are thus the characteristic exponents of the (vth order) system, and

$$Z_l = \prod_{\substack{s=1\\ s\neq l}}^{v}(\lambda_s I - \Gamma)\Big/\prod_{\substack{s=1\\ s\neq l}}^{v}(\lambda_s - \lambda_l).$$

Equation (3.33) can be recast as

$$\boldsymbol{x}_F(t) = P(t)\exp(\Gamma t)\int_0^t \sum_{k=-\infty}^{\infty}\sum_{l=1}^{v} P_k\exp\{[-\mu_l + \mathrm{j}(\omega_i + k\omega_p)]\tau\}\boldsymbol{Z}_l\boldsymbol{d}\mathrm{d}\tau.$$

The integration can now be performed readily by moving the integral inside the summation, viz.

$$\begin{aligned}
\boldsymbol{x}_F(t) &= P(t)\exp(\Gamma t)\boldsymbol{d}\sum_{k=-\infty}^{\infty}\sum_{l=1}^{v}P_k\boldsymbol{Z}_l\int_0^t \exp\{[-\mu_l + \mathrm{j}(\omega_i + k\omega_p)]\tau\}\mathrm{d}\tau \\
&= P(t)\exp(\Gamma t)\boldsymbol{d}\sum_{k=-\infty}^{\infty}\sum_{l=1}^{v}P_k\boldsymbol{Z}_l\frac{\exp[-\mu_l + \mathrm{j}(\omega_i + k\omega_p)]t - 1}{-\mu_l + \mathrm{j}(\omega_i + k\omega_p)} \qquad (3.34)\\
&= P(t)\exp(\Gamma t)\boldsymbol{d}\sum_{k=-\infty}^{\infty}\sum_{l=1}^{v}P_k\boldsymbol{Z}_l\frac{\exp[-\mu_l + \mathrm{j}(\omega_i + k\omega_p)]t}{-\mu_l + \mathrm{j}(\omega_i + k\omega_p)} - P(t)\exp(\Gamma t)h
\end{aligned}$$

with $h = \boldsymbol{d}\sum_{k=-\infty}^{\infty}\sum_{l=1}^{v}P_k\boldsymbol{Z}_l\frac{1}{-\mu_l + \mathrm{j}(\omega_i + k\omega_p)}$.

Using Sylvester's theorem to expand $\exp(\Gamma t)$, in the first term of Eq. (3.34), gives

$$\boldsymbol{x}_F(t) = P(t)\sum_{m=1}^{v}\exp(\mu_m t)\boldsymbol{Z}_m\boldsymbol{d}\sum_{k=-\infty}^{\infty}\sum_{l=1}^{v}P_k\boldsymbol{Z}_l\frac{\exp[-\mu_l + \mathrm{j}(\omega_i + k\omega_p)]t}{-\mu_l + \mathrm{j}(\omega_i + k\omega_p)}.$$

However since $Z_m Z_l = 0$ for all $m \neq l$, and $Z_m^2 = Z_m$[7] the above expression simplifies to

$$x_F(t) = P(t)\boldsymbol{d}\sum_{k=-\infty}^{\infty}\sum_{l=1}^{v}P_k Z_l\frac{\exp[\mathrm{j}(\omega_i + k\omega_p)t]}{-\mu_l + \mathrm{j}(\omega_i + k\omega_p)} - P(t)\exp(\Gamma t)h.$$

Expanding $P(t)$, in the left-hand term in a Fourier series according to

$$P(t) = \sum_{n=-\infty}^{\infty}P_n\exp(jn\omega_p t)$$

gives

$$\boldsymbol{x}_F(t) = \boldsymbol{d}\sum_{n=-\infty}^{\infty}\sum_{k=-\infty}^{\infty}\sum_{l=1}^{v}P_n P_k\frac{\exp[\mathrm{j}(\omega_i + q\omega_p)t]}{-\mu_l + \mathrm{j}(\omega_i + k\omega_p)} - P(t)\exp(\Gamma t)h\boldsymbol{Z}_l$$

where $q = n + k$.
Defining

$$v_{nk} = \boldsymbol{d}\sum_{l=1}^{v}\boldsymbol{Z}_l\frac{P_n P_k}{-\mu_l + \mathrm{j}(\omega_i + k\omega_p)}$$

then the complete forced response to a sinusoidal excitation can be expressed as

$$\boldsymbol{x}_F(t) = \sum_{n=-\infty}^{\infty}\sum_{k=-\infty}^{\infty}v_{nk}\exp[\mathrm{j}(\omega_i + q\omega_p)t] - P(t)\exp(\Gamma t)h. \qquad (3.35)$$

If the unforced system is stable—i.e. $\mathrm{Re}\{\mu_l\} < 0$ for all l (see Chap. 4) then $\exp(\Gamma t) \to 0$ as $t \to \infty$ showing that the right-hand term is a transient. If any $\mathrm{Re}\{\mu_l\} > 0$ then the right-hand term diverges to infinity. If $\mathrm{Re}\{\mu_l\} = 0$, for all l, then the right-hand term is periodic with period θ—i.e. the period of the parameter variation.

Most practical systems are stable so that the right-hand term does not appear after a sufficient time interval leaving as the steady state forced response of a general periodic system to a sinusoidal input

$$x_{ss}(t) = \sum_{n=-\infty}^{\infty} \sum_{k=-\infty}^{\infty} v_{nk} \exp\left[j(\omega_i + q\omega_p)t\right] \tag{3.36}$$

$$q = n + k.$$

A number of particularly important properties emerge from an examination of Eq. (3.36). For example the steady state response contains a large number of frequency components, including the input frequency but excluding the pump frequency, ω_p. Further the general steady state response is aperiodic unless ω_p and ω_i are commensurate. For example if $\omega_p = 2\omega_i$, the response will be periodic with fundamental frequency ω_i. The relative amplitudes of the various frequency components in the response will be determined by the coefficients v_{nk} which in turn will depend upon the relation between ω_p and ω_i and also upon the network or system, as embodied in the characteristic exponents μ_l.

Whether the steady state response, or indeed the complete forced response, can be evaluated analytically, depends, as with the natural solution, upon the availability of an analytical form for $\phi(t, \tau)$ as embodied directly in Eq. (3.29) and by way of its derived properties as in Eq. (3.32). Therefore periodic equations which lend themselves to an analytical determination of forced response are those whose coefficient waveforms render the homogeneous forms amenable to exact solution. Notwithstanding this, the above qualitative results are important guidelines in the interpretation of forced response in general.

3.15 Phase Space Analysis

Certain applications involving second order Hill equation descriptions do not require a complete record of a solution but rather demand a knowledge only of specific properties such as its global maximum. This is the case, for example, in quadrupole mass spectrometry [8]; here information of value can be derived from an examination of the solution in so-called *phase space*. Phase space is a two dimensional plot of the solution and its first derivative; at a particular time it appears simply as a point but with time describes a trajectory referred to as the phase space trajectory. Phase space trajectories for stable solutions to a Hill equation have some interesting properties as revealed in the detailed treatments of Hamilton [9] and Baril [10]. In this section, only the essential points are derived. In so far as quadrupole mass spectrometry is concerned the concept seems to have been introduced by Paul et al [11] but appears not to have been taken up until combined with matrix analytical techniques by Baril and Septier [12].

Let $u(t)$ and $v(t)$ be a pair of basis solutions to a Hill equation such that the complete solution at any time t can be expressed as

$$x(t) = au(t) + bv(t) \tag{3.37}$$

where a and b are constants to be determined. For a stable solution the basis functions, by association with Eq. (2.21), can be expanded as [1]

$$u(t) = \sum_{r=-\infty}^{\infty} C_r \cos(2r + \beta)t$$

$$v(t) = \sum_{r=-\infty}^{\infty} C_r \sin(2r + \beta)t$$

so that Eq. (3.37) can be written

$$x(t) = \sqrt{a^2 + b^2} \sum_{r=-\infty}^{\infty} C_r \cos[(2r + \beta)t + \phi] \tag{3.38}$$

with $\phi = \tan^{-1} b/a$. Values for a and b can be determined in the following way. By differentiating Eq. (3.37) it is seen that

$$\begin{bmatrix} x(t) \\ \dot{x}(t) \end{bmatrix} = \begin{bmatrix} u(t) & v(t) \\ \dot{u}(t) & \dot{v}(t) \end{bmatrix} \begin{bmatrix} a \\ b \end{bmatrix} = W(t) \begin{bmatrix} a \\ b \end{bmatrix}$$

so that

$$\begin{bmatrix} a \\ b \end{bmatrix} = W^{-1}(t) \begin{bmatrix} x(t) \\ \dot{x}(t) \end{bmatrix} = \frac{W^*(t)}{W_0} \begin{bmatrix} x(t) \\ \dot{x}(t) \end{bmatrix}$$

where $W(t)$ is the Wronskian matrix, $W^*(t)$ is its adjoint and W_0 is the Wronskian determinant. Values for a and b can be found by choosing a convenient time in the last expression, say $t = 0$, whereupon

$$a = [x(0)\dot{v}(0) - \dot{x}(0)v(0)]/W_0 \tag{3.39a}$$

and

$$b = [\dot{x}(0)u(0) - x(0)\dot{u}(0)]/W_0 \tag{3.39b}$$

Now, the maximum value that can be attained by the stable Hill equation solution is, from Eq. (3.38)

$$x_{\max} = \sqrt{a^2 + b^2} \sum_{r=-\infty}^{\infty} |C_r|$$

so that

$$x_{\max}^2 = (a^2 + b^2) \sum_{r=-\infty}^{\infty} |C_r|^2.$$

Substituting from Eq. (3.39a) and Eq. (3.39b) gives

$$\Gamma x(0)^2 + 2Ax(0)\dot{x}(0) + B\dot{x}(0)^2 = \varepsilon \tag{3.40}$$

which is the equation of an ellipse in phase space with

$$\Gamma = [\dot{v}(0)^2 + \dot{u}(0)^2]/W_0$$

$$A = [v(0)\dot{v}(0) + u(0)\dot{u}(0)]/W_0$$

$$B = [v(0)^2 + u(0)^2]/W_0$$

and

$$\varepsilon = x_{\max}^2 W_0 \Big/ \sum_{r=-\infty}^{\infty} |C_r|^2.$$

Thus solutions to the Hill equation with initial conditions lying on the ellipse of Eq. (3.40) will all have the same global bound, given that the solutions commenced at $t = 0$. For solutions commencing at a different time phase relative to the equation's periodic coefficient, the initial values of $u(t)$ and $v(t)$ and their derivatives will be different, leading to different ellipses in phase space, although with the same area.

From the definitions of $u(t)$ and $v(t)$ it is relatively straightforward to show that the solution at a time $m\pi$ (i.e. after m complete cycles of the coefficient) will fall on the initial condition ellipse although, in general, at a point different to that at the start. Moreover, if β can be expressed as a rational fraction, the trajectory in phase space, at the end of each complete cycle of the periodic coefficient, will fall only on a limited set of points on the appropriate ellipse and will exhibit a periodicity over those points.

The material above has most value when it is combined with an expression for the discrete state transition matrix, expressed in terms of the parameters of the ellipse. It is shown readily that [9]

$$\phi(\pi, 0) = \begin{bmatrix} \cos\beta\pi + A\sin\beta\pi & B\sin\beta\pi \\ -\Gamma\sin\beta\pi & \cos\beta\pi - A\sin\beta\pi \end{bmatrix}$$

whereupon it can be seen that

$$\det[\phi(\pi, 0)] = B\Gamma - A^2 = 1$$

3.16 Concluding Comments

It is evident in the foregoing that the ability to determine solutions to periodic differential equations rests upon being able to find a set of basis functions over a fundamental period. Therefore, in endeavouring to find a solution, sets of sub-intervals should be delineated over which linearly independent basis sets can be established. In some cases, such as with the Meissner equation, the set of sub-intervals is obvious. In other examples it may be necessary to ascribe intervals arbitrarily over which the coefficient can be closely approximated by a function (e.g. a ramp or step) for which basis solutions are known. This could be the case with a practically generated rectangular waveform, for which rise and decay segments, overshoot, preshoot etc. might be approximated by ramps. A detailed representation however is seldom required in practice. It is shown in Chap. 5 that the behaviour of periodically time-varying systems is more dependent upon the lower harmonic content of the periodic coefficient and is less sensitive to the higher harmonics. For an intractable equation then it might be more profitable to take advantage of this fact and model the periodic coefficient with some of the procedures discussed in Chap. 5 rather than trying graphically to represent the periodic function by combinations of steps, ramps and so on.

In some cases, nonlinear versions of Hill equations can be solved using the techniques just described. For example, if the periodic coefficient can be represented or modelled by a piecewise constant waveform, such as that discussed in Sect. 3.5, the equation in each time interval may appear as one of the standard constant coefficient nonlinear equations for which solutions can be deduced. The equation

$$\ddot{x} + \psi(t)(ax + bx^2 + cx^3 \cdots) = 0$$

for example, with $\psi(t)$ represented in piecewise constant form may have basis solutions on each time interval which are trigonometric, ellpitic or hyperelliptic functions depending upon the order of the polynomial nonlinearity.

Whilst the approach adopted above has been to derive state transition matrices by first ascertaining sets of basis functions and then deriving Wronskian matrices it may also be possible to make use of integral transforms for this task. This technique is commonly used in determining state transition matrices for constant coefficient systems. For example the equation

$$\ddot{x} + c^2 x = 0$$

written in matrix form is

$$\dot{\boldsymbol{x}}(t) = G\boldsymbol{x}(t) \tag{3.41}$$

in which $\boldsymbol{x}(t)$ is the vector $\begin{bmatrix} x(t) \\ \dot{x}(t) \end{bmatrix}$ and G is the constant matrix

$$G = \begin{bmatrix} 0 & 1 \\ -c^2 & 0 \end{bmatrix}. \tag{3.42}$$

The single sided Laplace transform of Eq. (3.41) is

$$s\bar{X}(s) - \boldsymbol{x}(0) = G\bar{X}(s)$$

where $\bar{X}(s)$ is the transform of $\boldsymbol{x}(t)$ and $\boldsymbol{x}(0)$ is a vector of initial conditions; s is the complex transform variable. Rearranging the last expression gives

$$\{sI - G\}\bar{X}(s) = \boldsymbol{x}(0)$$

so that

$$\bar{X}(s) = \{sI - G\}^{-1}\boldsymbol{x}(0)$$

whereupon taking the inverse transform gives

$$\boldsymbol{x}(t) = L^{-1}\{sI - G\}^{-1}\boldsymbol{x}(0) \tag{3.43}$$

with L^{-1} designating the inverse Laplace transform operator.

From the form of Eq. (3.43) the state transition matrix over the time interval $(0, t)$ is seen to be

$$\phi(t, 0) = L^{-1}\{sI - G\}^{-1}.$$

Using Eq. (3.42) this can be evaluated for the second order constant coefficient system being considered:

$$\begin{aligned}
\phi(t, 0) &= L^{-1}\begin{bmatrix} s & -1 \\ c^2 & s \end{bmatrix}^{-1} \\
&= L^{-1}\left\{\frac{1}{s^2+c^2}\begin{bmatrix} s & 1 \\ -c^2 & s \end{bmatrix}\right. \\
&= L^{-1}\begin{bmatrix} \dfrac{s}{s^2+c^2} & \dfrac{1}{s^2+c^2} \\ \dfrac{-c^2}{s^2+c^2} & \dfrac{s}{s^2+c^2} \end{bmatrix} \\
&= \begin{bmatrix} \cos ct & \frac{1}{c}\sin ct \\ -c\sin ct & \cos ct \end{bmatrix}
\end{aligned}$$

which should be compared to the state transition matrix derived in Sect. 3.2 for one interval of the rectangular waveform in the Meissner equation. The success of the Laplace transform in this derivation lies in the fact that the exponential kernel of the transformation is matched by the 'exponential' basis functions of the constant coefficient system. It is possible that other transformations could be used to deduce state transition matrices over intervals in which coefficients are time varying, provided again the kernels of those transforms match the basis function appropriate to the equation with that time variation.

References for Chapter 3

1. McLachlan, N. W.: Theory and application of Mathieu functions. Clarendon: Oxford U. P. 1947. Reprinted by Dover, New York 1964
2. Keenan, R. K.: An investigation of some problems in periodically parametric systems. PhD Thesis, Monash University Melbourne, 1966
3. Pipes, L. A.: Matrix solution of equations of the Mathieu-Hill type. J. Appl. Phys. 24 (1953) 902–910
4. Pipes, L. A.: Applied mathematics for scientists and engineers, 2nd ed. New York: McGraw-Hill 1958
5. McLachlan, N. W.: Bessel functions for engineers, 2nd ed. London: Oxford U. P. 1955
6. Abramowitz, M.; Stegun, I. A.: Handbook of mathematical functions. New York: Dover 1965
7. D'Angelo, H.: Linear time-varying systems: Analysis and sythesis. Boston: Allyn & Bacon 1970
8. Dawson, P: Quadrupole mass spectrometry. Amsterdam: Elsevier 1976
9. Hamilton, G. F.: The transition matrix for the Mathieu equation: Development and relation to u_{max}. Int. J. Mass Spectrom. Ion Phys. 28 (1978) 1–6
10. Baril, M: Etude des proprietes fondamentales de l'equation de Hill pour le dessin de filtre quadrupolaire. Int. J. Mass Spectrom. Ion Phys. 35 (1980) 179–200
11. Paul, W; Osberghaus, O; Fischer, E.: Ein Ionen-Käfig. Forschungsber. Wirtschafts-und Verkehrsministeriums Nordrhein-Westfalen: Koln, Opladen: Westdeutscher Verlag 1958
12. Baril, M; Septier, A: Piéreage des ions dans un champ quadrupolaire tridimensionnel a haute frequénce. Rev. Phys. Appl. 9 (1974) 525–531

Problems

3.1 The periodic coefficient in a Hill equation has the form shown below. Determine its discrete transition matrix.

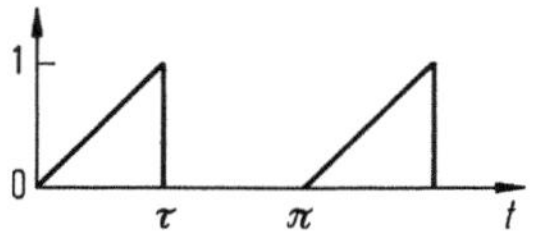

3.2 Demonstrate that the characteristic exponents for the Hill equations of Sects. 3.6 and 3.7 are identical (Hint: consider Eq. (2.27b)).

3.3 Show that $\det\{\phi(\pi, 0)\} = 1$ for the Meissner equation.

3.4 A Hill equation has the periodic coefficient shown below. Sketch the form of solution over two periods of the coefficient. Repeat this exercise for a coefficient that is the negative of that shown.

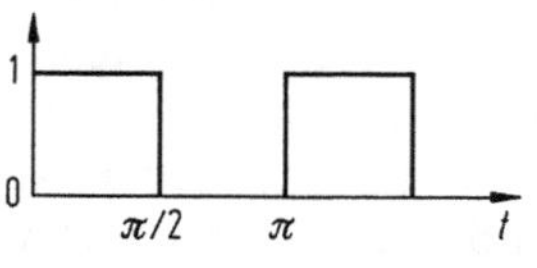

3.5 The periodic coefficient in a particular Hill equation consists of a repetitive sequence of decaying exponentials. Deduce an expression for the corresponding discrete transition matrix and for the Wronskian. Is it possible to use this to deduce the discrete transition matrix for a Hill equation with an impulsive coefficient?

3.6 In Sect. 3.2 the basis functions for either of Eq. (3.3a) or Eq. (3.3b) were chosen as $\cos ct$ and $\frac{1}{c}\sin ct$ where $c^2 = a \pm 2q$ as appropriate. What effect would the choice of simply $\sin ct$ for the second basis function have on the expression for the discrete transition matrix in Eq. (3.5a)?

3.7 Consider the third order Meissner equation

$$\frac{d^3x}{dt^3} + (a - 2q\psi(t))x = 0$$

where $\psi(t)$ is the rectangular coefficient of Fig. 3.1. Determine expressions (in principle) for the Wronksian matrices over the intervals $(0, \tau)$ and (τ, π) and thus indicate how a computer program could be written to evaluate the discrete transition matrix for a range of values of a and q.

3.8 Determine the state transition matrix for a Hill equation with a periodic coefficient in the form of a sequence of uniformly spaced, alternating positive and negative impulses.

Chapter 4

Stability

The stability of the solution to a periodically time-varying equation can depend critically upon very small changes in its parameters. Moreover, unlike constant coefficient equations, the dependence of stability upon a particular parameter can be complicated in that there are often ranges of parameter values for which a periodic system is unstable, separated by regions of stability. As a result, the problem of stability of periodically time-varying systems has received detailed attention in the past, especially for systems of second order. Additionally, many applications involving Hill equations rely principally upon stability and to a lesser extent upon the actual forms of solution, thereby adding to the interest shown in this aspect of parametric behaviour.

4.1 Types of Stability

From a practical point of view there are two broad categories of stability that should be distinguished. These are *classical stability* and *short time stability*. Briefly, a system is regarded as classically stable if its response remains bounded as time goes to infinity. On the other hand a system is said to be short time stable if its response remains within a certain specified bound during a given time of observation. Clearly a system can be classically stable, yet short time unstable, and vice versa. This is depicted in Fig. 4.1.

In practice, short time stability is all that can be observed, although of course if an observation can be made over a long enough time and the system response can be measured, then classical stability can be implied by the 'short time' stability observed. Conversely, although practical systems should be treated in terms of short time stability considerations, classical stability is often a sufficiently accurate substitute. There are some exceptions to this in practice wherein the use of classical stability can lead to inaccurate results; in general however, parametric systems are described almost entirely in terms of classical stability, not the least reason for which is the difficulty in determining short time stability conditions. Conditions of these types are discussed in Sect. 4.6. The intervening sections are addressed exclusively to classical stability so that *unstable* and *stable* are used in that sense.

There are many definitions used for stability. Comprehensive treatments can be found in Chap. 1 of Willems [1] and Chap. 8 of D'Angelo [2]. Some definitions refer to the boundedness of forced response—if the output of a system is bounded

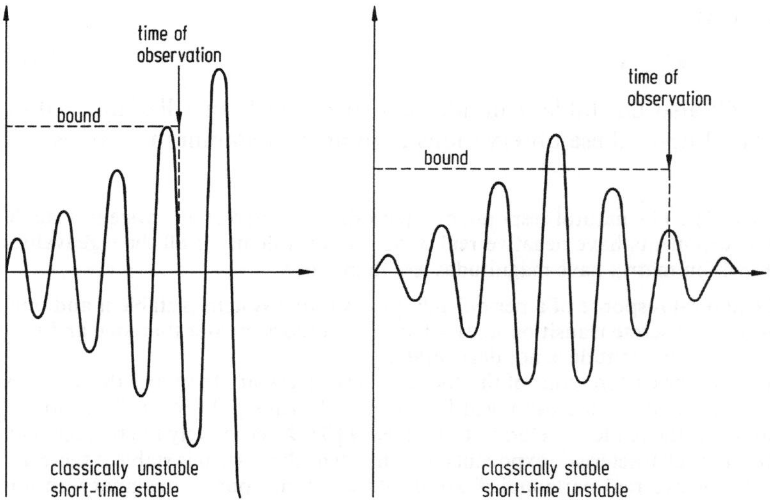

Fig. 4.1. Illustration of the concept of short time stability. Note that classically stable solutions can be short time unstable, and vice versa

for a bounded input the system is said to be *non-resonant* or *stable bounded-input, bounded output*. Conversely a system which is *resonant* is one for which the output grows without bound for an input of finite amplitude. The lossless *LC* circuit is, of course, the textbook example of resonance. It is important to realise that stability and non-resonance are not equivalent. The *LC* network for example is stable but resonant. Resonance is covered in detail by D'Angelo and, since it does not arise in the examples of periodic differential equations to be treated in this book, will not be considered further.

Some authors define a system as stable if the response of the unforced system, to arbitrary initial conditions, approaches zero as time goes to infinity. This is a restrictive definition, particularly for practical applications, for which reason the following more general definition of stability is used in this book. A system is said to be stable if the natural response remains bounded as time goes to infinity. This is strictly referred to as *Lagrange stability* or *global asymptotic stability* or simply *asymptotic stability*. A system which is not Lagrange stable is referred to as *unstable*.

4.2 Stability Theorems for Periodic Systems

Stability tests for periodically time-varying systems are based, either explicitly or implicitly, upon an examination of the nature of the characteristic exponents or characteristic multipliers (eigenvalues of the discrete transition matrix) of Eq. (2.22). For an unforced periodic system to be stable it is sufficient that

$$\mathrm{Re}\{\mu_i\} < 0 \qquad \forall i \tag{4.1a}$$

which is the same as

$$|\lambda_i| < 1 \qquad \forall i. \tag{4.1b}$$

Such a system will also be stable if in addition to Eq. (4.1) one $\mathrm{Re}\{\mu_i\} = 0$ or one $|\lambda_i| = 1$. Based upon these observations a number of stability theorems can be stated.

Theorem 4.1 (Eqs. (4.1)). The natural response of a periodically time-varying system is stable if all characteristic exponents have negative real parts, or equivalently if all the eigenvalues of the discrete transition matrix have magnitudes less than unity.

Theorem 4.2. The natural response of a periodically time-varying system is stable if and only if no eigenvalues of the discrete transition matrix have magnitudes greater than one and that an eigenvalue with unity magnitude is not degenerate.

Theorem 4.2 can also be cast in terms of the Jordan blocks necessary to reduce the matrices C and Γ in Sec. 2.4 to their Jordan canonical forms—see Willems [1]. For a discussion of Jordan canonical forms the reader is referred to Faddeeva [3]. Alternatively the theorem can be expressed in terms of characteristic exponents in as much as the system is stable if there are no exponents with positive real parts and if exponents which are purely imaginary do not form complex conjugate pairs of the form $\pm j2\frac{k\pi}{\theta}$ where k is an integer. This can be deduced from Eq. (2.22) and the statement of Theorem 4.2 above.

The above theorems can be restated as conditions for instability, as in the following.

Theorem 4.3. The natural response of a periodically time-varying system is unstable if

(i) there is at least one characteristic exponent with a positive real part, or equivalently at least one eigenvalue of the discrete state transition matrix of magnitude greater than unity, or

(ii) there is at least one eigenvalue of the discrete transition matrix with unit magnitude and multiplicity greater than unity.

The significance of part (ii) of theorem 4.3, and thus also of theorem 4.2, is that a situation of multiple eigenvalues will lead to a solution of the t-multiplied variety. Thus the solution will contain a term of the form $(1 + t)\lambda$ or equivalently $e^{\mu t}(1 + t)$. When $|\lambda| < 1$ or $\mathrm{Re}\{\mu\} < 0$, the t in the coefficient will not increase as quickly as the exponential decrease with time thus ensuring stability. However for $\mathrm{Re}\{\mu\} = 0$ or $|\lambda| = 1$, the $(1 + t)$ coefficient will cause divergence of the solution and thus instability. Unfortunately the importance of this has not been sufficiently emphasised in standard treatments of the Hill equation, yet it is precisely this situation of degenerate eigenvalues with unit magnitudes that generates the particular cases of instability of importance in applications such as parametric amplification.

With respect to the stability of the forced response of a periodic system, it will be assumed here that the forced solution will be stable if the natural response is stable, with the exception of a situation of resonance in a lossless system as discussed previously.

4.3 Second Order Systems

4.3.1 Stability and the Characteristic Exponent

Equation (2.27a) permits a straightforward analysis of the stability of a lossless second order periodic system, once the discrete transition matrix $\phi(\theta, 0)$ is known.

Theorem 4.2 above shows that a necessary condition for stability is that $|\lambda| < 1$. Applying this to Eq. (2.27a) reveals that

$$|\text{trace}\{\phi(\theta, 0)\}| < 2 \tag{4.2}$$

for a system to be stable. Under this condition the eigenvalues of the discrete transition matrix form a complex conjugate pair, given by

$$\lambda_{1,2} = T/2 \pm \sqrt{1 - (T/2)^2}$$

with $T = \text{trace}\{\phi(\theta, 0)\}$. When the trace of the discrete transition matrix is identically equal to ± 2, the eigenvalues become real and degenerate of unit magnitude, leading to t-multiplied instability.

Should the value of the characteristic exponent be desired it can be computed from Eq. (2.27b).

For a second order system with losses Eq. (2.26a) can be used, after the elements of $\phi(\theta, 0)$ have been determined. The condition for stability then generalises to

$$|\text{trace}\{\phi(\theta, 0)\}| < \det\{\phi(\theta, 0)\} + 1 \tag{4.3a}$$

and

$$\det\{\phi(\theta, 0)\} \leq 1. \tag{4.3b}$$

Alternatively, for a lossy system, it is necessary to find the value of the characteristic exponent μ of the equation when transformed to its lossless counterpart using Eq. (2.4), whereupon the system will be stable if

$$\text{Re}\{\mu\} < \frac{1}{2}\int_0^\theta g_1(t)\mathrm{d}t \tag{4.4}$$

where $g_1(t)$ is the coefficient of the first derivative (loss) term, as seen in Eq. (2.3). The actual characteristic exponent for the lossy system is

$$\mu' = \mu - \frac{1}{2}\int_0^\theta g_1(t)\mathrm{d}t \tag{4.5}$$

from which it is seen that $\text{Re}\{\mu'\} < 0$ if Eq. (4.4) applies.

4.3.2 The Meissner Equation

The Meissner equation is a lossless Hill equation with a rectangular waveform coefficient. Its discrete transition matrix and solution have been treated in Sect. 3.2 to 3.4. It is perhaps the most readily handled Hill equation and is used here to illustrate stability.

For a rectangular waveform coefficient as illustrated in Fig. 3.1, in which τ is the length of the positive segment and π the period of the coefficient, the discrete state transition matrix is given by Eq. (3.5a). From this, the trace is easily shown to be

$$\text{trace}\{\phi(\pi, 0)\} = 2\cos d(\pi - \tau)\cos c\tau - \left[\frac{d}{c} + \frac{c}{d}\right]\sin d(\pi - \tau)\sin c\tau \tag{4.6}$$

where $c^2 = a - 2q$ and $d^2 = a + 2q$. In view of Eq. (4.2) a necessary and sufficient condition for the stability of the Meissner equation is that

$$\left|\cos d(\pi - \tau)\cos c\tau - \frac{1}{2}\left[\frac{d}{c} + \frac{c}{d}\right]\sin d(\pi - \tau)\sin c\tau\right| < 1. \tag{4.7}$$

which is easily computed for given a, q and τ.

It is instructive now to consider several illustrations from Eq. (4.7) since they will highlight the complexity of the stability of the Meissner equation and thus of periodic differential equations in general.

Consider $\tau = \pi/2$ so that the coefficient reduces to a square waveform.

If in addition $q = 0$ then $c = d = \sqrt{a}$ and Eq. (4.7), for a square waveform coefficient, becomes

$$|\cos^2 \tfrac{1}{2}\pi\sqrt{a} - \sin^2 \tfrac{1}{2}\pi\sqrt{a}| < 1$$

i.e.

$$|\cos \pi\sqrt{a}| < 1.$$

For a positive the inequality is satisfied and thus the system is stable, although only marginally so when $\sqrt{a} = n$, n integer. In other words at the specific values

$$a = n^2 \tag{4.8}$$

the system will have a pair of real eigenvalues of value unity and thus will exhibit t-multiplied instability at those points.

For a negative the inequality above is clearly not satisfied and the second order lossless Meissner equation is always unstable for $q = 0$.

As another illustration consider q non-zero but a zero, still with a square waveform coefficient. Under these circumstances $c = \pm \mathrm{j}\sqrt{2q}$, $d = \pm\sqrt{2q}$ and Eq. (4.7) becomes

$$|\cos \tfrac{1}{2}\pi\sqrt{2q}\cosh \tfrac{1}{2}\pi\sqrt{2q}| < 1.$$

Clearly as q increases from zero the chance that the inequality will be satisfied diminishes owing to the increase in the magnitude of the hyperbolic cosine term. Moreover, owing to the cyclic nature of the ordinary cosine term there will be points of stability corresponding to $\sqrt{2q} = 2n + 1$ and, depending upon the magnitude of $\cosh \frac{1}{2}\pi\sqrt{2q}$, there will be ranges of q about those points for which stability prevails. These ranges reduce in size as q increases. It is clear that substituting $q = -q$ will leave the inequality unchanged and thus the stability will be identical for negative q. This implies simply that stability of the Meissner equation, for the special cases of a square wave coefficient and $a = 0$, is unaffected by inverting (and thus shifting by one half a period) the periodic coefficient.

The stability of the Meissner equation is clearly very complex in its dependence upon the equation parameters a and q (not to mention τ which has been held at $\pi/2$ in the above examples). It is conveniently summarised however by constructing a map of values of a and q on which stability or instability, as appropriate, is indicated for each (a, q) pair. Such a map is referred to as a *stability diagram*, and is shown for the Meissner equation in Fig. 4.2, for the special case of a square waveform coefficient.

The shaded areas of the diagram represent values of a and q for which the

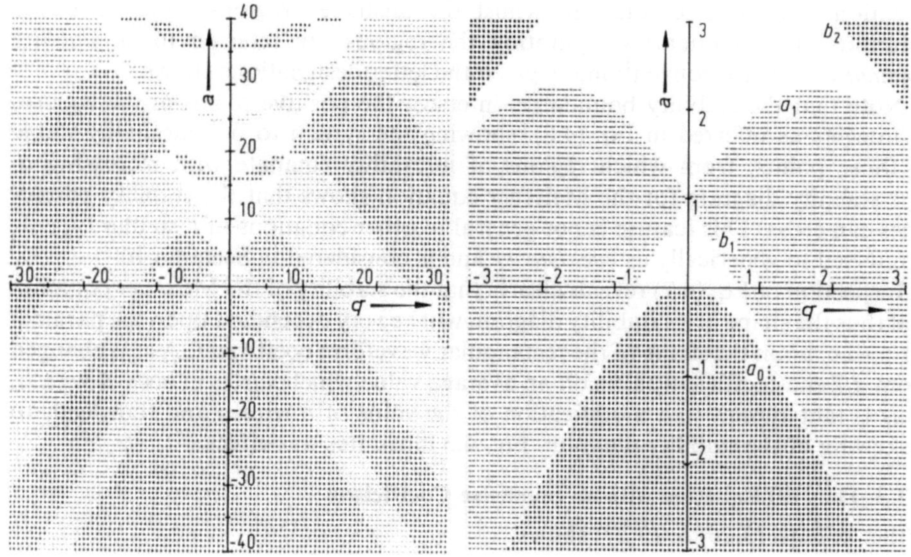

Fig. 4.2. Stability diagram for the Meissner equation ($\tau = \pi/2$). Blank regions represent values of a and q that give stable solutions. The other regions give unstable solutions. The diagram was produced computationally by evaluating the trace of the discrete transition matrix according to Eqs. (4.6) and (4.7)

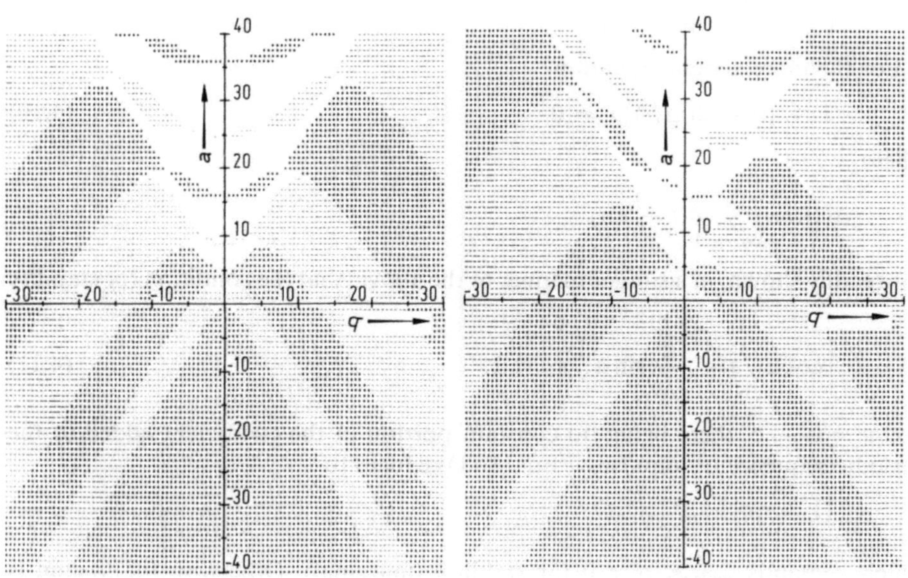

Fig. 4.3. Stability diagrams for the Meissner equation showing the effect of varying τ, the positive porch length in the rectangular coefficient. Blank regions give stable solutions

equation has unstable solutions whilst the unshaded regions represent stable solutions. The boundaries separating the regions are referred to as *stability boundaries* and are conventionally given the letter designations shown.

Note that the stability boundaries intersect the positive a axis at the squares of integers as inferred in Eq. (4.8), showing the system to be marginally stable at those points, from which regions of instability emanate with increasing q. The stability diagram for the Meissner equation shows that the stability boundaries can cross. This feature is not general to other equations of a similar type. It is referred to generically as *coexistence* and is considered in Sect. 4.5.10.

Inspection of Eq. (4.7) reveals clearly that the stability of the Meissner equation and thus the shape of its stability diagram will be very dependent upon the parameter τ—i.e. upon the shape of the rectangular waveform coefficient. As τ is changed from $\pi/2$ the coefficient takes on an average value which varies in accord with q. This is tantamount to a modification of the value of a with q and thus leads to a skewing of the stability diagram. Fig. 4.3 illustrates this effect.

4.3.3 The Hill Equation with an Impulsive Coefficient

Either of Eqs. (3.26) or (3.27) shows

$$\text{trace}\,\{\phi(\pi, 0)\} = 2\cos d\pi - \frac{1}{d}\sin d\pi \tag{4.9}$$

for the Hill equation with periodic impulsive coefficient of the type shown in Fig. 3.7. From Eq. (4.2) such an equation will therefore be stable whenever

$$\left| 2\cos d\pi - \frac{1}{d}\sin d\pi \right| < 2.$$

For the more general equation

$$\ddot{x} + \{a - 2q\Delta(t)\}x = 0 \tag{4.10}$$

where

$$\Delta(t) = \sum_{k=-\infty}^{\infty} \delta(t - k\pi)$$

it is readily shown, in a similar manner to the above, that the condition for stability is

$$\left| \cos \pi\sqrt{a} + \frac{q}{\sqrt{a}}\sin \pi\sqrt{a} \right| < 1. \tag{4.11}$$

For $a = 0$—i.e. there being no constant portion of the coefficient added to the periodic sequence of impulses—Eq. (4.11) reduces to

$$|1 + \pi q| < 1$$

which demands $-\frac{2}{\pi} \leq q < 0$ for stability (marginal). Otherwise the system is always unstable for $a = 0$. This can be seen in the complete stability diagram for Eq. (4.10) shown in Fig. 4.4.

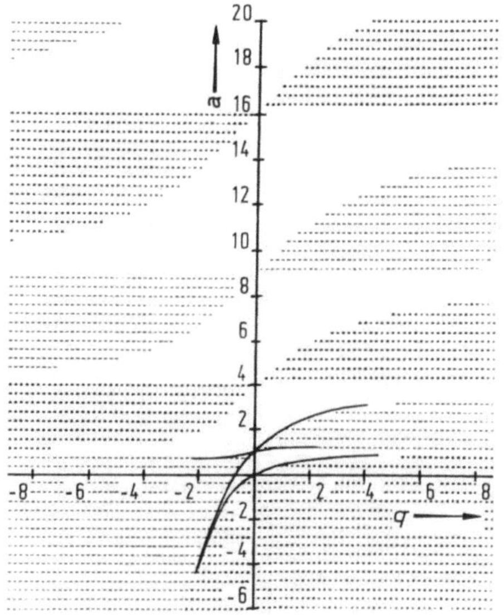

Fig. 4.4. Stability diagram for the Hill equation with an impulsive coefficient. Blank regions give stable solutions

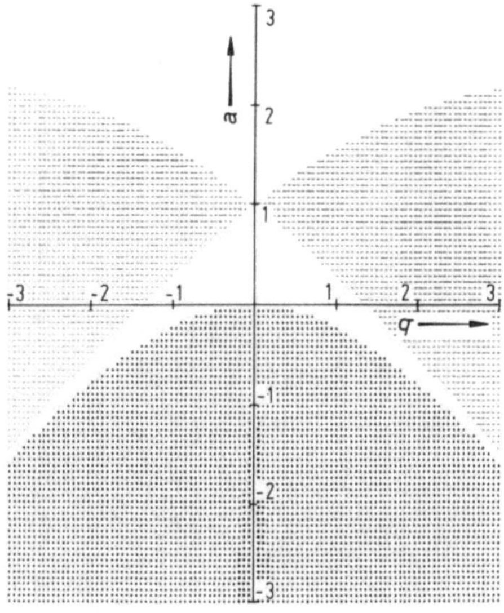

Fig. 4.5. Stability diagram for a Hill equation with either a positive or negative slope sawtooth waveform coefficient. Blank regions give stable solutions

4.3.4 The Hill Equation with a Sawtooth Waveform Coefficient

Inspection of Eqs. (3.22a) and (3.23a) reveals that the traces of the discrete transition matrices for Hill equations with triangular waveform coefficients of positive and negative slope, are the same.

Accordingly, in view of Eq. (4.2) stability, as would be expected, is independent of whether the sawtooth is positive or negative. Again, using Eq. (4.2) along with the trace expression permits a stability diagram to be generated as in Fig. 4.5.

4.3.5 The Hill Equation with a Triangular Waveform Coefficient

The stability of a Hill equation possessing a triangular coefficient such as that represented in Fig. 3.5 will depend upon the parameter τ. Again, using the information provided in Sect. 3.8 it is a relatively easy task to assess stability and to produce stability diagrams. Shown in Fig. 4.6 are diagrams for a symmetric triangle and a sawtooth coefficient ($\tau = \pi/2$ and 0 respectively).

4.3.6 Hill Determinant Analysis

If the discrete state transition matrix can be found for a particular second order periodic system then assessment of stability, using the various expressions presented above, is fairly straightforward. Unfortunately, this is very often the exception, with most periodic differential equations eluding analysis. A notable example is the Mathieu equation. For these difficult equations there are two methods for obtaining usable stability checks. One is to use the modelling tech-

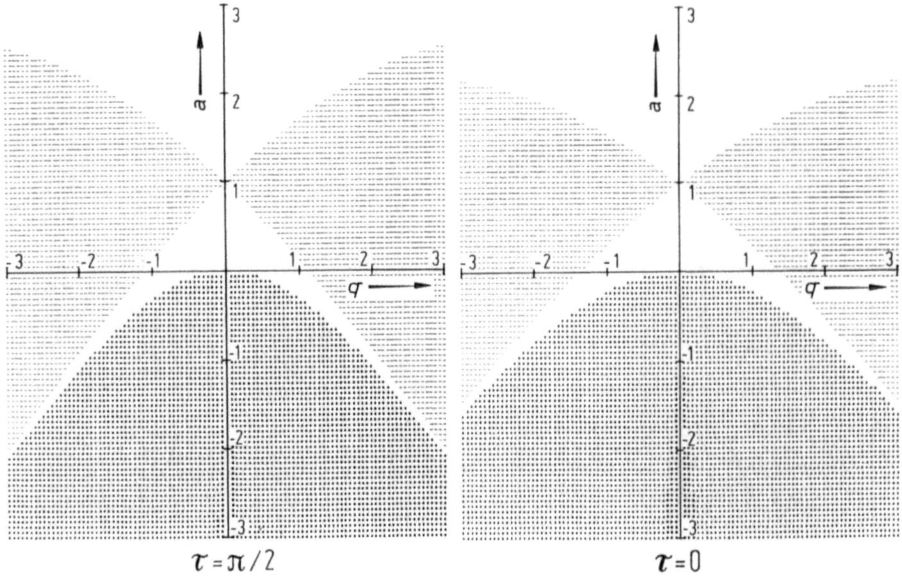

Fig. 4.6. Stability diagrams for a Hill equation with a triangular waveform coefficient showing the influence of waveform asymmetry. When $\tau = \pi/2$ the waveform is perfectly symmetrical whereas for $\tau = 0$ it degenerates to a sawtooth waveform. Blank regions represent stable solutions

niques of Chap. 5. The other is to use the approach known generally as Hill determinants, this being a classical (although sometimes cumbersome) method for determining stability.

Analysis using Hill determinants has been a popular technique for studying periodic differential equations since its inception in 1877. It rests upon the assumed Floquet solution (see Eq. (2.21)) of the form

$$x(t) = e^{\mu t}\Xi(t), \quad \Xi(t) = \Xi(t + \pi), \tag{4.12}$$

for the general second order equation

$$\ddot{x} + (a - 2q\psi(t))x = 0, \quad \psi(t) = \psi(t + \pi).$$

Owing to its periodicity $\psi(t)$ can be expressed in a complex Fourier series of the form

$$\psi(t) = \sum_{n=-\infty}^{\infty} \Psi_n \exp(jn\omega_p t) \tag{4.13}$$

in which the Ψ_n are the complex Fourier coefficients and ω_p is the pumping frequency, of value

$$\omega_p = \frac{2\pi}{\theta} = 2[\text{since } \theta = \pi \text{ (by convention)}]. \tag{4.14}$$

For generality it is retained as ω_p in the following. With Eq. (4.13) the general second order Hill equation can be written

$$\ddot{x} + x \sum_{n=-\infty}^{\infty} G_n \exp(jn\omega_p t) = 0 \tag{4.15}$$

in which

$$G_0 = a - 2q\Psi_0 \tag{4.16a}$$

and

$$G_n = -2q\Psi_n. \tag{4.16b}$$

The function $\Xi(t)$ in Eq. (4.12) can also be expressed in a Fourier series

$$\Xi(t) = \sum_{r=-\infty}^{\infty} C_r \exp(jr\omega_p t)$$

which, when substituted in Eq. (4.15) and dividing throughout by $e^{\mu t}$, yields

$$\sum_{r=-\infty}^{\infty} \left\{ C_r(\mu + jr\omega_p)^2 \exp(jr\omega_p t) + \sum_{n=-\infty}^{\infty} G_n C_r \exp[(n + r)j\omega_p t] \right\} = 0.$$

Equating coefficients of $\exp(jm\omega_p t)$ to zero, where m is an integer, leaves the infinite set of simultaneous equations

$$(\mu + jr\omega_p)^2 C_r + \sum_{n=-\infty}^{\infty} G_n C_{r-n} = 0 \tag{4.17}$$

where r takes on all values within the interval $(-\infty, \infty)$. Equation (4.17) is a set of homogeneous equations in the unknown C_r. (μ is also unknown, of course, however methods for finding its value will emerge.) In order that the set have a nontrivial solution it is necessary that the determinant of its coefficients be singular (i.e. zero). The convergence[1] of this determinant will be secured if the equations are divided throughout by

$$\rho_r = (\mu + jr\omega_p)^2 + G_0 \tag{4.18}$$

leaving Eq. (4.17) as

$$C_r + \sum_{\substack{n=-\infty \\ n \neq 0}}^{\infty} G_n \rho_r^{-1} C_{r-n} = 0. \tag{4.19}$$

---

[1] Convergence of infinite determinants is discussed in McLachlan [4].

The resulting infinite determinant of coefficients is therefore

$$\Delta(\mu) = \begin{vmatrix} & \vdots & \vdots & \vdots & \vdots & & \\ \cdots & 1 & \frac{G_1}{\rho_{-2}} & \frac{G_2}{\rho_{-2}} & \frac{G_3}{\rho_{-2}} & \frac{G_4}{\rho_{-2}} & \cdots \\ \cdots & \frac{G_{-1}}{\rho_{-1}} & 1 & \frac{G_1}{\rho_{-1}} & \frac{G_2}{\rho_{-1}} & \frac{G_3}{\rho_{-1}} & \cdots \\ \cdots & \frac{G_{-2}}{\rho_0} & \frac{G_{-1}}{\rho_0} & 1 & \frac{G_1}{\rho_0} & \frac{G_2}{\rho_0} & \cdots \\ \cdots & \frac{G_{-3}}{\rho_1} & \frac{G_{-2}}{\rho_1} & \frac{G_{-1}}{\rho_1} & 1 & \frac{G_1}{\rho_1} & \cdots \\ \cdots & \frac{G_{-4}}{\rho_2} & \frac{G_{-3}}{\rho_2} & \frac{G_{-2}}{\rho_2} & \frac{G_{-1}}{\rho_2} & 1 & \cdots \\ & \vdots & \vdots & \vdots & \vdots & \vdots & \end{vmatrix}. \tag{4.20}$$

$$= 0, \; \rho_r \neq 0.$$

In principle solution of Eq. (4.20) will yield a value for the characteristic exponent provided the G_n are known, which they generally are in any practical situation. Obtaining the roots of $\Delta(\mu)$ however is a difficult problem and, instead of attempting this, the determinantal equation is reduced to a more readily handled trigonometric form, as below. The procedure followed is that used by McLachlan [4]. In Sect. 4.4.1 an alternative approach is presented.

The determinant $\Delta(\mu)$ has simple poles, corresponding to the zeros of

$$\rho_r = (\mu + \mathrm{j}r\omega_p)^2 + G_0 = 0. \tag{4.21}$$

These are its only singularities. In the μ plane they appear as

$$\mu = \pm\mathrm{j}\sqrt{G_0} - \mathrm{j}r\omega_p. \tag{4.22}$$

Now the function

$$\chi(\mu) = [\cos \mathrm{j}2\pi\mu/\omega_p - \cos 2\pi\sqrt{G_0}/\omega_p]^{-1} \tag{4.23}$$

has the same singularities as $\Delta(\mu)$ [2] and, in addition has the same periodicity with μ. Thus the function

$$\lambda(\mu) = \Delta(\mu) - C\chi(\mu)$$

will have no singularities if C is suitably chosen. A function which has no singularities is, by Liouville's theorem, a constant. To determine $\lambda(\mu)$ and thus also C we note that as $\mu \to \infty$ all $\rho_r \to 0$ leaving $\Delta(\infty) = |\boldsymbol{I}|$ the determinant of the

[2] Assuming $G_0 \neq (r\omega_p)^2$, so that $\mu = 0$ is not a pole.

identity matrix, which of course is unity. Also $\lim_{\mu \to \infty} \chi(\mu) = 0$ so that $\lambda(\mu) = 1$ giving

$$C = [\Delta(\mu) - 1]/\chi(\mu).$$

Since C is a constant it can be evaluated for any suitably chosen value of μ. If $\mu = 0$, $\chi(0) = [1 - \cos 2\pi\sqrt{G_0}/\omega_p]^{-1}$ and thus

$$C = [\Delta(0) - 1][1 - \cos 2\pi\sqrt{G_0}/\omega_p]$$

where $\Delta(0)$ is $\Delta(\mu)$ evaluated at $\mu = 0$. Note that for given values of a and q (the parameters in the periodic coefficient) and for $\psi(t)$ defined explicitly, $\Delta(0)$ can be evaluated, at least for finite approximations of the doubly infinite form.

Now

$$\frac{1}{\chi(\mu)} = C/[\Delta(\mu) - 1]$$

which, using the above value for C, becomes

$$\frac{1}{\chi(\mu)} = [\Delta(0) - 1][1 - \cos 2\pi\sqrt{G_0}/\omega_p]/[\Delta(\mu) - 1].$$

However from Eq. (4.23)

$$\frac{1}{\chi(\mu)} = \cos j2\pi\mu/\omega_p - \cos 2\pi\sqrt{G_0}/\omega_p$$

so that we have

$$\cos j2\pi\mu/\omega_p - \cos 2\pi\sqrt{G_0}/\omega_p = [\Delta(0) - 1][1 - \cos 2\pi\sqrt{G_0}/\omega_p]/[\Delta(\mu) - 1].$$

Now Eq. (4.20) gives $\Delta(\mu) = 0$ as the condition necessary for determining the μ (and thus also the C_r of Eq. (4.17) should they be required). Putting this in the last expression gives

$$\begin{aligned} \cos j2\pi\mu/\omega_p &= [1 - \Delta(0)][1 - \cos 2\pi\sqrt{G_0}/\omega_p] + \cos 2\pi\sqrt{G_0}/\omega_p \\ &= 1 - \Delta(0)[1 - \cos 2\pi\sqrt{G_0}/\omega_p] \end{aligned}$$

i.e.

$$\cos j2\pi\mu/\omega_p = 1 - 2\Delta(0)\sin^2 \pi\sqrt{G_0}/\omega_p. \tag{4.24}$$

Equation (4.24) is a particularly useful expression in the analysis of stability of a second order Hill equation and also in the determination of values for the characteristic exponent μ.

Normally the pumping period in second order systems is chosen as π (i.e. $\omega_p = 2\,\text{rad}\cdot\text{s}^{-1}$) so that Eq. (4.24) can be recast more simply as

$$\cos j\pi\mu = 1 - 2\Delta(0)\sin^2 \frac{\pi}{2}\sqrt{G_0}. \tag{4.25}$$

If the doubly infinite determinant $\Delta(0)$ is suitably truncated to a manageable size, say 3×3 or 5×5, Eq. (4.25) can be used to advantage to approximate stability and to generate rough stability diagrams for Hill equations. Numerical evaluation, making use of larger finite truncations of $\Delta(0)$ gives more accurate results but is time consuming.

4.3.7 Parametric Frequencies for Second Order Systems

A notable common feature of the stability diagrams presented in Figs. 4.2 to 4.6 is that regions of instability converge on the a axis at values of a given by

$$a = n^2. \tag{4.26}$$

This is a general result for second order periodically pumped systems and can be proved using Eq. (4.25) in the following manner.

As $q \to 0$, Eq. (4.20) shows $\Delta(0) \to 1$. Also $G_0 \to a$ giving from Eq. (4.25)

$$\cos \mathrm{j}\pi\mu = 1 - 2\sin^2\frac{\pi}{2}\sqrt{a}$$
$$= \cos \pi\sqrt{a}.$$

For a negative the above expression shows μ to be real, demonstrating the system to be unstable as observed in all second order diagrams. However for a positive

$$|\cos \mathrm{j}\pi\mu| \leq 1$$

with the equality holding only when $a = n^2$. Consequently μ is always imaginary and, for $a = n^2$ has the form

$$\mu = \pm \mathrm{j}k, \quad k\text{-integer}$$

which, according to theorem 4.2, is a condition for (marginal) instability.

It is instructive from a practical viewpoint to reinterpret Eq. (4.26) in terms of system parameters. For this consider a second order system with pumping, expressed as

$$\ddot{x} + (a' - 2q'\psi(t))x = 0$$

with $\psi(t) = \psi(t + \theta)$. The equivalent static (unpumped) system is

$$\ddot{x} + a'x = 0$$

which has roots of $\pm \mathrm{j}\sqrt{a'}$. Thus the natural frequency of the static system is, as is well known,

$$\omega_0 = \sqrt{a'}.$$

Now, to make use of Eq. (4.26) it is necessary to transform the time variable to make the pumping period π. Consequently, introduce the canonical time

$$\xi = \omega_p t/2 = \pi t/\theta,$$

with which the pumped equation becomes

$$\ddot{x} + \frac{4}{\omega_p^2}.(a' - 2q'\psi(\xi))x = 0$$

where the dots above x now refer to differentiation with respect to ξ. If we define

$$a = \frac{4a'}{\omega_p^2}$$

$$q = \frac{4q'}{\omega_p^2}$$

then the last equation becomes

$$\ddot{x} + (a - 2q\psi(\xi))x = 0$$

which is in the standard form of a second order Hill equation, with a pumping period of π. For such an equation Eq. (4.26) shows instability, for small pumping, to occur at

$$a = n^2, \quad n\text{-integer},$$

which thus shows

$$\frac{4a'}{\omega_p^2} = n^2$$

to be a condition for instability (and, as will be seen in Chap. 7, a condition for parametric gain). Since $a' = \omega_0^2$ where ω_0 is the natural frequency of the static system, then the condition for instability and gain, for small pump magnitudes, is

$$\omega_p = \frac{2\omega_0}{n}. \tag{4.27}$$

Thus if a second order system has a parameter pumped at twice the system natural frequency, or at integral submultiples of that rate, instability and gain is likely.

For a lossless system instability will be observed for every pump frequency given by Eq. (4.27). However when losses are present pump amplitudes have to be increased to invoke instability and further, it is found that more pumping is required as n is increased. This effect is seen clearly in Fig. 6.5.

4.4 General Order Systems

Owing to the added complexities introduced very little worthwhile theory has been generated for periodic systems of general order, again apart from special cases such as those involving rectangular waveform coefficients. Those are readily treated using the Wronskian and state transition matrix techniques developed in Sects. 2.2 to 2.4. It is possible however to lay the foundations of a Hill determinant procedure for equations of general order and, from this, say something about parametric frequencies. Beyond this, recourse is perhaps best had to numerical methods.

4.4.1 Hill Determinant Analysis for General Order Systems

The general Hill equation of Eq. (2.1) can be expressed as

$$\sum_{k=0}^{v} [\alpha_k + q\gamma_k\psi(t)]\overset{k.}{x} = 0. \tag{4.28}$$

in which the expressions for the coefficients have been separated into their constant (α_k) and time varying ($q\gamma_k\psi(t)$) parts. In Eq. (4.28) it is assumed that $|\psi(t)|_{\max} = 1$ and $\psi(t) = \psi(t + \theta)$. The factors γ_k are scaling constants introduced by the topology of the system and have the effect of weighting the time variation across the appropriate terms in the equation. For any derivatives without an associated time-varying coefficient the corresponding γ_k will be zero. The variable q, as with second order systems, expresses the degree of pumping the (overall) system is subjected to. There is no term equivalent to a in second order systems. Instead it is replaced by the set α_k. For simplicity it is assumed that $\psi(t)$ has no average value, although this is readily included if necessary. It's inclusion will not affect the ultimate results here—i.e. of determining parametric frequencies for periodically time-varying systems in general.

The analysis now proceeds very much as in Sect. 4.3.6 although it is more complicated owing to the additional terms present. Also it is not possible, as will be seen, to invoke Liouville's Theorem with the same facility, thus limiting the scope of the present analysis.

A fundamental (Floquet) solution to Eq. (4.28) is

$$x_i(t) = \exp(\mu_i t) \sum_{r=-\infty}^{\infty} C_r \exp(\mathrm{j} r\omega_p t) \tag{4.29}$$

in which μ_i is the characteristic exponent associated with that solution and the C_r are a set of coefficients (to be determined, desirably). Also $\omega_p = 2\pi/\theta$. Further $\psi(t)$ may be expanded as

$$\psi(t) = \sum_{n=-\infty}^{\infty} \Psi_n \exp(\mathrm{j} m\omega_p t),\ \Psi_0 = 0 \tag{4.30}$$

so that substituting Eqs. (4.29) and (4.30) into Eq. (4.28) gives, after equating coefficients of $\exp(\mathrm{j} l\omega_p t)$, l integral,

$$\sum_{k=0}^{v} \left\{ C_r \alpha_k (\mu_i + \mathrm{j} r\omega_p)^k + q\gamma_k \sum_{n=-\infty}^{\infty} \Psi_n C_{r-n} \cdot [\mu_i + \mathrm{j}(r-n)\omega_p]^k \right\} = 0, \quad -\infty < r < \infty. \tag{4.31}$$

Defining

$$\rho_r = \sum_{k=0}^{v} \alpha_k (\mu_i + \mathrm{j} r\omega_p)^k \tag{4.32}$$

and

$$H_{n,r} = q\Psi_n \sum_{k=0}^{v} \gamma_k [\mu_i + \mathrm{j}(r-n)\omega_p]^k \tag{4.33}$$

Eq. (4.31) may be written, after dividing throughout by ρ_r, as

$$C_r + \rho_r^{-1} \sum_{\substack{n=-\infty \\ n\neq 0}}^{\infty} C_{r-n} H_{n,r} = 0 \quad -\infty < r < \infty \tag{4.34}$$

which, again, is a set of homogeneous equations in the infinity of unknown C_r. In order that the set possess a nontrivial solution it is necessary that the determinant of coefficients be singular. Thus

$$\Delta(\mu) = \begin{vmatrix} \ddots & & & & \\ 1 & & & & \\ \dfrac{H_{-1,-1}}{\rho_{-1}} & 1 & \dfrac{H_{1,-1}}{\rho_{-1}} & \dfrac{H_{2,-1}}{\rho_{-1}} & \\ \dfrac{H_{-2,0}}{\rho_0} & \dfrac{H_{-1,0}}{\rho_0} & 1 & \dfrac{H_{1,0}}{\rho_0} & \\ & \dfrac{H_{-2,1}}{\rho_1} & \dfrac{H_{-1,1}}{\rho_1} & 1 & \\ & & & \ddots & \end{vmatrix} = 0 \tag{4.35}$$

As with second order systems, the above expression is particularly difficult to use analytically as a result of which it is desirable to replace it by an equivalent algebraic form. To this end note that the only singularities of $\Delta(\mu)$ in μ are the zeros of ρ_r, given by

$$\rho_r = \sum_{k=0}^{v} \alpha_k y^k = 0 \tag{4.36}$$

where $y \equiv \mu_i + jr\omega_p$. Equation (4.36) is a polynomial equation in y which will be recognised as the characteristic (or indicial) equation of Eq. (4.28) for $q = 0$. If the roots of Eq. (4.36) are designated Y_i, $i = 1, \ldots, v$, then

$$\rho_r = \prod_{i=1}^{v} (y - Y_i) = \prod_{i=1}^{v} (\mu_i + jr\omega_p - Y_i). \tag{4.37}$$

Owing to the general nature of the Y_i it is not possible to invent a suitable function along the lines of that in Eq. (4.23) for second order equations. Consequently the development now follows a direction different to that for second order Hill equations.

It is convenient to assume the Y_i are distinct (although essentially the same results will be obtained when degenerate Y_i exist), as a result of which $\Delta(\mu)$ has only simple poles in μ. Moreover $\Delta(\mu)$ is bounded as μ approaches infinity so that the determinant may be expressed [5]

$$\Delta(\mu) = C + \sum_{r=-\infty}^{\infty} \sum_{i=1}^{v} A_{ir}(\mu_i + jr\omega_p - Y_i)^{-1} \tag{4.38}$$

where C is a constant and the A_{ir} are the residues at the simple poles. Letting $\mu_i \to \infty$ in Eq. (4.38) causes the principal parts to tend to zero and $\Delta(\mu_i) \to 1$ so that $C = 1$.

Now Eq. (4.38) shows $\Delta(\mu)$ is to be periodic in μ_i with a period of θ/r. As a result it is assumed, following Meadows [6] that the residues of $\Delta(\mu)$ at its poles are independent of the integer r so that

$$A_{i1} = A_{i2} = \cdots = A_{iv} \stackrel{\text{def}}{=} A_i.$$

On reversing the order of summation, Eq. (4.38) can therefore be written as

$$\Delta(\mu) = 1 + \frac{1}{j\omega_p}\sum_{i=1}^{v} A_i \sum_{r=-\infty}^{\infty}\left\{r + (\mu_i - Y_i)\frac{1}{j\omega_p}\right\}^{-1}. \tag{4.39}$$

However since

$$\sum_{r=-\infty}^{\infty}(r + z)^{-1} = \pi \cot \pi z$$

then Eq. (4.39) becomes

$$\Delta(\mu) = 1 + \frac{\theta}{2}\sum_{i=1}^{v} A_i \coth\{(\mu_i - Y_i)\theta/2\}.$$

Thus for $\Delta(\mu) = 0$, as required in Eq. (4.35), we have

$$\sum_{i=1}^{v} A_i \coth\{(\mu_i - Y_i)\theta/2\} = -2/\theta. \tag{4.40}$$

Equation (4.40) is now a generalised algebraic counterpart to the determinantal expression of Eq. (4.35), from which values of the characteristic exponents μ_i can be generated. To proceed to this generally however requires an evaluation of the residues A_i, which themselves are expressed as convergent, doubly infinite determinants as will be seen in the following.

4.4.2 Residues of the Hill Determinant for $q \to 0$.

The residue of $\Delta(\mu)$ at the pole $\mu_i = Y_i - jr\omega_p$ is given in the usual way as

$$A_i = \lim_{\mu_i \to Y_i - jr\omega_p} (\mu_i + jr\omega_p - Y_i)\Delta(\mu)\big|_{\mu_i = Y_i - jr\omega_p}.$$

Thus A_i will be a determinant with zeros on its principal diagonal. Moreover because of the form of $\Delta(\mu)$ all the nonzero terms of A_i will have q as a factor. As a result of these points A_i will have q as a factor and thus approaches zero with q.

Now Eq. (4.40) may be written

$$\sum_{i=1}^{v} A_i \frac{\cosh\{(\mu_i - Y_i)\theta/2\}}{\sinh\{(\mu_i - Y_i)\theta/2\}} = -2/\theta$$

whereupon multiplying throughout by

$$\prod_{i=1}^{v} \sinh\{(\mu_i - Y_i)\theta/2\}$$

and letting $q \to 0$ gives

$$\prod_{i=1}^{v} \sinh\{(\mu_i - Y_i)\theta/2\} = 0.$$

Since the zeros of a product can be given by examining the zeros of the individual terms, the following expression emerges for determining the characteristic exponents for $q \to 0$ and thus for assessing the stability of a periodic system for small pumping:

$$\sinh\{(\mu_i - Y_i)\theta/2\} = 0$$

which thus gives

$$\mu_i = Y_i + \mathrm{j}l\omega_p \tag{4.41}$$

where l is an integer. In view of Eq. (2.22), for small pumping amplitudes the eigenvalues for a general periodic system are thus given by

$$\lambda_i = \exp(Y_i\theta). \tag{4.42}$$

4.4.3 Instability and Parametric Frequencies for General Systems

For a real, passive system the roots Y_i will all have real parts equal to or less than zero; from Eq. (4.42) it is seen therefore that the corresponding eigenvalues will have magnitudes less than unity. Consequently such a system is always stable for small pumping. For a system with a purely imaginary root the corresponding eigenvalue will have a unit magnitude and, by itself, cannot contribute to instability. If however there exist a pair of degenerate eigenvalues, associated with a pair of purely imaginary Y_i the system will be unstable for small pumping according to theorem 4.2 in Sect. 4.2. For a passive system imaginary roots will exist in complex conjugate pairs and thus if

$$Y_{i,j} = \pm\mathrm{j}\beta$$

the corresponding eigenvalues, of unit magnitude are

$$\lambda_{i,j} = \exp(\pm\mathrm{j}\beta\theta).$$

These will only be degenerate if

$$\theta = n\pi/\beta, \tag{4.43}$$

where n is an integer, or, in terms of pump frequency, when the system parameters are pumped at the rate

$$\omega_p = 2\beta/n. \tag{4.44}$$

Since β is an imaginary root it is one of the system natural frequencies. With this in mind Eq. (4.44) should be compared with Eq. (4.27). Whereas the latter suggests that instability (and parametric gain) can be most easily invoked in a second order system by pumping at $2/n$ times the natural frequency, Eq. (4.44) says that instability and gain can be readily produced in a general order system by pumping at $2/n$ times *any* of the natural frequencies.

4.4.4 Stability Diagrams for General Order Systems

Since the parameter a is not defined for systems of general order the concept of an (a, q) stability diagram has no meaning. However reflection on the meaning of a for a second order system according to the analysis of Sect. 4.3.7 reveals that variation of a amounts equivalently to changing the system natural frequency or the pumping frequency. Whilst the former is acceptable for second order systems it is not unique when the order is four and higher. However variation of pumping frequency, for a given system, can be entertained regardless of order. Consequently for systems of general order it is sensible to plot a stability diagram with axes which describe amplitude of pumping (via q) and frequency of pumping, ω_p. Rather than use ω_p directly it is preferable to employ θ since Eq. (4.43) shows that this will display more clearly the essential results for differing n. The most important practical region in a diagram is that in the vicinity of $n = 1$, so that if frequency were plotted, the diagram would display severe compression near the origin with increasing n.

As an illustration of the nature of θ versus q stability diagrams for high order systems, Fig. 4.7 shows an example—for a fourth order rectangularly-pumped system with the static pole positions indicated. These are the poles of the unpumped system and thus reflect the natural frequencies. Their real parts give an indication of the losses in the system. In particular if the real part associated with a pole is small it is easy to invoke gain (or instability) with small pump magnitudes by pumping at the corresponding pump frequencies given by Eq. (4.44). Conversely large pumping is required if the real parts of the poles are large. This feature can be clearly seen in the figure.

The stability diagram of Fig. 4.7 shows a tendency to instability and gain at a set of frequencies not given by Eq. (4.44). These are designated as 'c' type regions in the figure. It is readily determined that these pump frequencies are given approximately by $(\beta_1 + \beta_2)/n$, where β_1 and β_2 are the natural frequencies of the static system; they are thus *coupled parametric modes* of the type described (for $n = 1$) by Sato [7].

4.5 Natural Modes and Mode Diagrams

It is seen in Chap. 2 for a vth order periodic system that there are v basis or fundamental solutions and thus v characteristic exponents, or equivalently v eigenvalues of the discrete transition matrix. The collection of fundamental solutions for a particular equation and a particular set of parameters is referred to as a *mode* of that equation. It is found that only specific sets of modes exist and these are readily described in terms of the nature of the eigenvalues of the discrete transition matrix. This section, which follows closely although not in such detail the treatment of Keenan[8], develops the essential characteristics of the modes of a periodic system and therefrom presents the concept of a mode diagram.

4.5.1 Nature of the Basis Solutions

Equation (2.21) shows that the basis solutions of a periodic differential equation can be expressed in the Floquet form:

$$x_i(t) = \exp(\mu_i t)\Xi_i(t) \tag{4.45}$$

in which μ_i is the (complex) characteristic exponent and $\Xi_i(t)$ is a bounded function

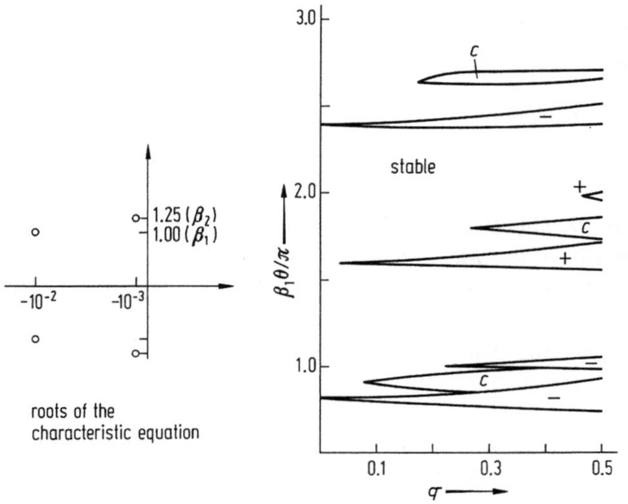

Fig. 4.7. Stability diagram for a fourth order periodic system with the static pole positions indicated. The parameters are varying in a rectangular fashion. (IEEE Trans. Circuits Syst. CAS-24 (1977) 241–247)

of period θ. The basis solution takes on a different nature according to the form of μ_i and thus of λ_i, as expressed in Eq. (2.22). Keenan's classification of the fundamental solutions in fact depends upon the type of λ_i and this will be followed here. These eigenvalues can be positive or negative real or complex conjugate. The function $\Xi_i(t)$ echoes the character of the corresponding eigenvalue. Keenan shows for positive or negative real eigenvalues $\Xi_i(t)$ is real and periodic with period θ, whereas for complex eigenvalues $\Xi_i(t)$ is complex and still periodic of period θ. Further since complex eigenvalues occur in conjugate pairs, so do the $\Xi_i(t)$.

4.5.2 *P* Type Solutions

When an eigenvalue of the discrete transition matrix is positive real the corresponding basis solution given in Eq. (4.45) is called *P* type. From Eq. (2.22) the characteristic exponent is seen to be of the form

$$\mu_i = \alpha_i + \mathrm{j}\frac{2\pi k}{\theta}, \quad k\text{-integer} \tag{4.46}$$

whilst as noted above $\Xi_i(t)$ is also real. Consequently the *P* type solution is of the form

$$\exp(\alpha_i t)\exp\left(\mathrm{j}\frac{2\pi k}{\theta}t\right)\Xi_i(t) \tag{4.47}$$

$$\stackrel{\text{def}}{=} \exp(\alpha_i t)\phi(t).$$

Now

$$\begin{aligned}\phi(t+\theta) &= \exp\left(\mathrm{j}\frac{2\pi}{\theta}t\right)\exp\left(\mathrm{j}\frac{2\pi\theta}{\theta}\right)\Xi_i(t+\theta)\\ &= \exp\left(\mathrm{j}\frac{2\pi t}{\theta}\right)\Xi_i(t)\\ &= \phi(t).\end{aligned}$$

Thus $\phi(t)$ is periodic of period θ. Therefore the P type solution is an exponentially modulated, oscillatory function, of pseudo-period θ. Equation (2.23) shows there are no constraints on the number of P type fundamental solutions for a particular equation. It will be induced in Sect. 4.5.4 however that at least one P type solution must occur for odd ordered periodic systems.

4.5.3 *C* Type Solutions

For a complex eigenvalue λ_i, Eq. (2.22) shows that the characteristic exponent is, in general, also complex. The corresponding C type solution is therefore expressible as

$$\exp(\alpha_i + \mathrm{j}\beta_i t)\Xi_i(t) \tag{4.48}$$

which is an exponentially modulated product of a pair of periodic functions, one of frequency β_i and the other of frequency ω_p (i.e. period θ). In some cases (e.g. second order lossless systems) $\alpha_i = 0$ so that the single C type solution appears as the product of two periodic functions. However since complex λ_i and $\Xi_i(t)$ have conjugate partners, C type solutions always exist in complex conjugate pairs in the form (neglecting α_i)

$$\exp(\mathrm{j}\beta_i t)\Xi_i(t) + \exp(-\mathrm{j}\beta_i t)\Xi_i^*(t)$$

which is a real function exhibiting both amplitude and phase modulation. This can easily be written

$$2|\Xi_i(t)|\cos(\beta_i t + \underline{/\Xi_i(t)}) \tag{4.49}$$

in which the modulations are more apparent.

4.5.4 *N* Type Solutions

N type solutions arise from real negative eigenvalues of the discrete transition matrix. From Eq. (2.22) the corresponding characteristic exponent must be of the form

$$\mu_i = \alpha_i + \mathrm{j}(2k+1)\pi/\theta, \quad k\text{-integer}.$$

Choosing the simple case of $k = 0$ the complete N solution is

$$\exp\alpha_i t\exp(\mathrm{j}\pi t/\theta)\Xi_i(t) \overset{\mathrm{def}}{=} \exp\alpha_i t f(t). \tag{4.50}$$

The function $f(t)$ in this N type solution is particularly interesting. Note that

$$\begin{aligned}f(t+\theta) &= \exp\mathrm{j}\pi t/\theta\exp\mathrm{j}\pi\Xi_i(t+\theta)\\ &= -\exp\mathrm{j}\pi t/\theta\Xi_i(t)\\ &= -f(t)\end{aligned}$$

and that

$$f(t + 2\theta) = f(t).$$

Thus N solutions are periodic with a period of twice that of the parameter variation. In other words N type solutions are subharmonic to the pump frequency. This is an important point since many practical illustrations of parametric behaviour depend upon invoking N type solutions. Again note that, as with P type solutions, N solutions have an impressed exponential modulation.

Equation (2.23) shows that negative eigenvalues must exist in pairs to preserve an overall positive product. Consequently N type solutions exist in pairs. A direct outcome of this is for odd ordered systems, there must be at least one P type solution.

4.5.5 Modes of Solution

The collection of all fundamental solutions for a particular equation is, as noted earlier, referred to as a mode of the equation for its particular set of parameters. Different modes of solution will exist as the parameters in the equation are altered, and from a practical view, especially as the amplitude and rate of pumping are varied. Consequently it is useful to have available a summary of the possible modes for a given equation so that the nature of its solutions may be assessed. Such a summary is the *mode diagram*; this depicts the modes of solution that will result for different pump frequencies and amplitudes. It is very similar in appearance to the stability diagram discussed earlier and indeed, for lossless second order systems, their boundaries are identical. In the following subsections some illustrative mode diagrams are presented and used to say something of parametric behaviour.

4.5.6 The Modes of a Second Order Periodic System

Equation (2.24a) shows immediately for a lossless second order system, the only modes possible are those of two P type solutions, two N type solutions or two C type solutions. These are referred to respectively at $2P$, $2N$ and $2C$ modes. Further Eq. (2.27a) shows that the conditions for existence of the various modes are

$$2P\colon \operatorname{trace}\{\phi(\theta, 0)\} > 2 \tag{4.51a}$$

$$2C\colon |\operatorname{trace}\{\phi(\theta, 0)\}| < 2 \tag{4.51b}$$

$$2N\colon \operatorname{trace}\{\phi(\theta, 0)\} < -2. \tag{4.51c}$$

Comparing these conditions with those for stability given in Sect. 4.3 it is readily seen that, for lossless second order equations, $2C$ modes are always stable whereas $2P$ and $2N$ modes are always unstable. Moreover it can be shown by a simple computation that regions of instability alternate as $2N$, $2P$, $2N$ etc. modes as one moves up through the stability diagram, as illustrated in Fig. 4.8 for the Meissner equation. An important practical feature to be extracted from this diagram is that in pumping according to Eq. (4.27) to achieve gain for low pump amplitudes, a response subharmonic to the pump is given for $a = 1, 9, 25$ etc.

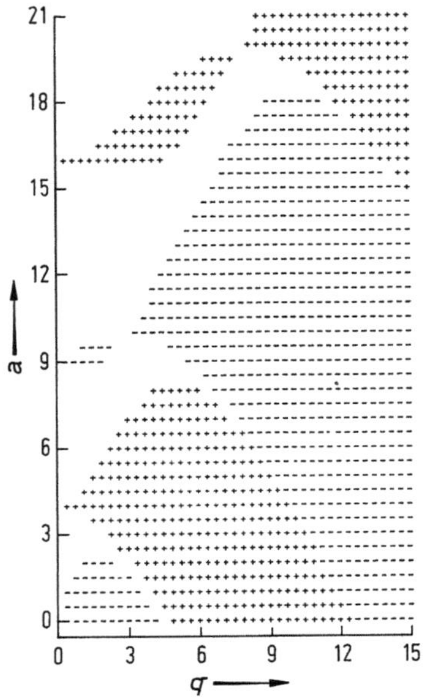

Fig. 4.8. Combined mode and stability diagram for the Meissner equation. Blank (2C) regions are stable whereas the others (2N and 2P) are unstable

(N type solutions) whilst a response at the pump frequency is generated with a = 4, 16, etc. (P type solutions).

4.5.7 Boundary Modes

Equations (4.51) show that the system modes are presently undefined for the trace of the discrete transition matrix being identically equal to two—in other words when the eigenvalues are degenerate. Since complex eigenvalues can only exist in conjugate pairs, degenerate eigenvalues must be either positive or negative real. The corresponding P or N type fundamental solutions will therefore be degenerate and no longer a linearly independent pair. Instead new solutions form of a t-multiplied variety giving modes with solution types of

$$P, tP \qquad (B_1)$$

and

$$N, tN. \qquad (B_2)$$

Following Keenan these are designated B_1 and B_2 type modes respectively. It follows that the conditions for these modes are therefore

$$B_1: \text{trace}\{\phi(\theta, 0)\} = 2$$

$$B_2: \text{trace}\{\phi(\theta, 0)\} = -2.$$

Clearly these modes will exist on the boundaries separating other types of mode. Keenan also refers to them as Brillouin modes.

In general a condition for the existence of boundary modes (and thus mode boundaries in mode diagrams) is that a pair of eigenvalues be degenerate. This is for a system of any order.

4.5.8 Second Order System with Losses

When losses are present in a second order periodic system a first derivative appears in the corresponding differential equation. As with Eq. (2.3) the first derivative term can be removed with the transformation of Eq. (2.4). The modes of the original system which results will be those of the transformed system modified by the transformation function

$$\exp\left\{-\frac{1}{2}\int_0^t g_1(t)\mathrm{d}t\right\}$$

where $g_1(t)$ is the coefficient of the first derivative term. If $g_1(t)$ is a constant, the transformation simply becomes

$$\exp\{-g_1 t/2\}$$

which thus does not alter the types of mode for any given parameters but rather only affects stability. The change in stability is illustrated for the lossy Mathieu equation in Sect. 6.4.

If $g_1(t)$ is not constant but varying periodically at the pump frequency then the transformation function above will be of the form

$$\exp(-\alpha_0 t)\gamma(t)$$

with $\gamma(t) = \gamma(t + \theta)$. Therefore, again the modes will not be changed. Only stability will be affected, depending upon α_0[8].

4.5.9 Modes for Systems of General Order

As with stability it is difficult to determine the modes of a general periodically time-varying system, unless an explicit form for the discrete transition matrix can be found. The best that can be done in general is to use the fact that mode boundaries are defined by degenerate eigenvalues and endeavour to approximate conditions for degenerate eigenvalues. As an example of this Eq. (4.42) gives an approximate expression for the eigenvalues of the discrete transition matrix for a general order system when the pump magnitude is small, i.e. $q \to 0$. In that equation the Y_i are the roots of the static (umpumped) system. A pair of eigenvalues will be degenerate if a pair of complex conjugate roots exist, of the form

$$Y_{i,K} = \alpha \pm \mathrm{j}\beta$$

with

$$\beta = k\pi/\theta.$$

In other words, a condition for mode boundaries is that

$$\theta = k\pi/\beta$$

which is the same as that given in Eq. (4.43) for instability with the significant

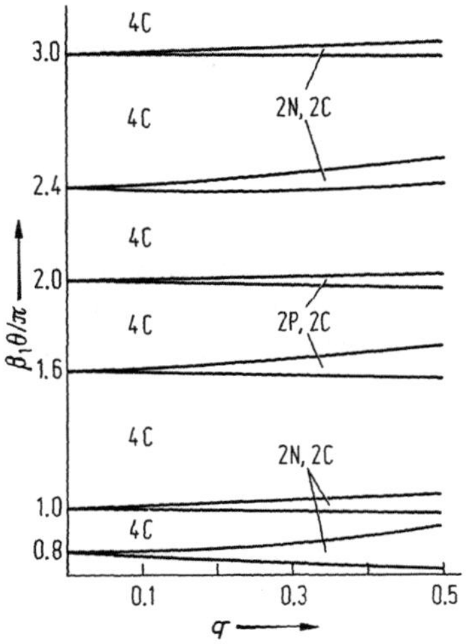

Fig. 4.9. Mode diagram for the fourth order system whose stability diagram appears in Fig. 4.7. (IEEE Trans. Circuits Syst. CAS-24 (1977) 241–247)

difference that for mode boundaries the real part of the system static roots is unimportant.

In Fig. 4.9 is shown the mode diagram for the system whose stability diagram is illustrated in Fig. 4.7. Again θ and q have been (logically) chosen as axes. It is seen that mode boundaries emanate from the θ axis for any degree of loss whereas stability boundaries only approach the θ axis for small damping.

4.5.10 Coexistence

Figures 4.2, 4.3 and 4.8 show that the stability boundaries for the Meissner equation can cross each other for non-zero values of q. The same feature can be observed for some other Hill equations but does not occur in the special case of the Mathieu equation, except for the trivial situation of $q = 0$, where stability boundary intersections occur at values of a that are squares of integers. Figure 4.3 demonstrates that the intersections, if present, take place between boundaries referred to as a-type and those designated b-type. In Sect. 6.1.1 following it is shown, in the particular case of the Mathieu equation, that on these boundaries the equation has one purely periodic solution. (The other solution is a t-multiplied version). On a-type boundaries it is an even function whereas on b-type boundaries it is odd. The same description is true of the Meissner equation in general. Thus stability boundary intersections of the type observed for the Meissner equation imply *coexistence*, at special values of a and q, of an odd and an even periodic function, both of the same period (since they border common regions of stability).

The question of coexistence of odd and even periodic solutions for non zero q has been addressed by Arscott [9], who presents a theorem by Ince which demonstrates that coexistence cannot occur for the Mathieu equation specifically. Hochstadt [10] has treated coexistence of the Meissner equation as have Hiller and Keenan [11] whereas Keenan [8] has considered the situation for the Hill equation in general.

4.6 Short Time Stability

The concept of short time stability has been described in Sect. 4.1 and illustrated in Fig. 4.1. It is seen to be related to a solution or system response remaining below a certain bound over a given time of observation. In this section the concept is considered in more detail and sufficient conditions for assessing the short time stability of periodically time-varying systems are presented. Proofs of these conditions are not given but may be found in D'Angelo [2] and Dorato [12].

Consider a time-varying, linear homogeneous system expressed in standard vector form

$$\dot{\boldsymbol{x}}(t) = G(t)\boldsymbol{x}(t) \tag{4.52}$$

where $\boldsymbol{x}(t)$ is the solution or state vector and $G(t)$ is a coefficient matrix, which in the case of a periodic system has period θ. This system is said to be short time stable if

$$\boldsymbol{x}^t(0)\boldsymbol{x}(0) \leq \varepsilon$$

implies

$$\boldsymbol{x}^t(t)\boldsymbol{x}(t) \leq C$$

over the time interval $[0, T]$. T is the time of observation whereas ε relates to the upper bound on the system's initial state and C describes the bound below which its state at any time must fall. The superscript t denotes transpose.

Assessment of short time stability is made in terms of the symmetric matrix [2]

$$U(t) = \tfrac{1}{2}\{G(t) + G^t(t)\}. \tag{4.53}$$

Sufficient conditions for short term stability of the system are:

Condition 1.

$$\int_0^t \lambda_M(\tau)\mathrm{d}\tau \leq \frac{1}{2}\ln\frac{C}{\varepsilon},\ 0 \leq t \leq T \tag{4.54}$$

where $\lambda_M(t)$ is the largest eigenvalue of $U(t)$, for each t.

OR

Condition 2. The principal minors of

$$M(t) = \left\{\frac{1}{2T}\ln\frac{C}{\varepsilon}\cdot I - U(t)\right\} \tag{4.55}$$

be equal to or greater than zero over the period $[0, T]$, where I is the identity matrix.

OR

Condition 3.

$$\int_0^T \|U(\tau)\|\mathrm{d}\tau \leq \frac{1}{2}\ln\frac{C}{\varepsilon} \tag{4.56}$$

where $\|U(t)\|$ is the matrix norm defined by

$$\|U(t)\| = \sum_{i,j} |u_{ij}(t)|.$$

It is evident that there are similarities in Eqs. (4.54) and (4.56). In fact it is known that $\lambda_M(t) \le \|U(t)\|$ so that Eq. (4.56) is much coarser than Eq. (4.54) and would therefore only be of consideration when the eigenvalue $\lambda_M(t)$ is difficult to determine. Generally Eq. (4.55) is also coarser than Eq. (4.54) but again is usually easier to apply, involving the evaluation of minor determinants rather than eigenvalues.

For a second order Hill equation $G(t)$ has the form (see Sect. 2.5)

$$G(t) = \begin{bmatrix} 0 & 1 \\ -(a - 2q\psi(t)) & 0 \end{bmatrix}, \psi(t) = \psi(t + \pi)$$

so that

$$U(t) = \begin{bmatrix} 0 & \dfrac{1-a}{2} + q\psi(t) \\ \dfrac{1-a}{2} + q\psi(t) & 0 \end{bmatrix}. \tag{4.57}$$

Therefore

$$\|U(t)\| = |1 - a + 2|q||\psi(t)|| \tag{4.58}$$

and

$$M(t) = \begin{bmatrix} \dfrac{1}{2T}\ln\dfrac{C}{\varepsilon} & \dfrac{a-1}{2} - q\psi(t) \\ \dfrac{a-1}{2} - q\psi(t) & \dfrac{1}{2T}\ln\dfrac{C}{\varepsilon} \end{bmatrix}. \tag{4.59}$$

The three sufficient conditions for the short time stability of a second order Hill equation can therefore be expressed as follows.

Condition 1. The eigenvalues of $U(t)$ are $\pm\left\{\dfrac{1-a}{2} + q\psi(t)\right\}$. Consequently, the 'largest' eigenvalue is

$$\lambda_M(t) = \frac{1-a}{2} + |q||\psi(t)| \quad \text{for } a < 1$$

$$= \frac{a-1}{2} + |q||\psi(t)| \quad \text{for } a > 1,$$

and

$$= |q||\psi(t)| \quad \text{for } a = 1.$$

Condition 1 becomes therefore,

$$|q|\int_0^t |\psi(\tau)|d\tau \pm \left(\frac{1-a}{2}\right)t \le \frac{1}{2}\ln\frac{C}{\varepsilon} \quad \text{for } a \lessgtr 1.$$

$$0 \le t \le T.$$

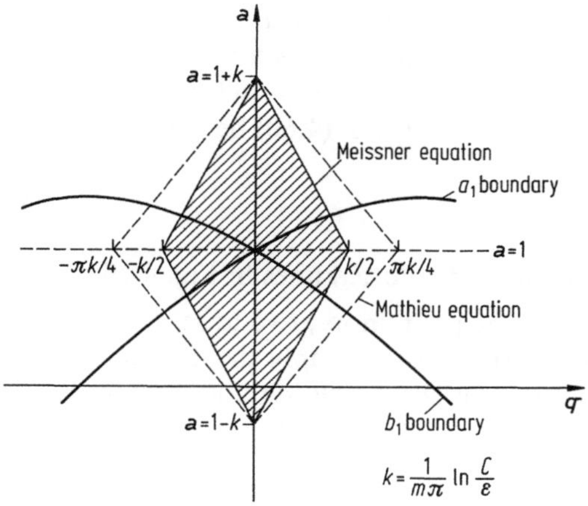

Fig. 4.10 Regions of short time stability for the Meissner and Mathieu equations. The equations have short time stable solutions within the boxes indicated

Owing to the absolute nature of the integrand the upper limit can be set to T. Also in the second term on the left hand side $t = T$ since the term is monotonically increasing. Furthermore, it is convenient to consider short time stability over an integral number of periods, say m so that $T = m\pi$. As a result the above expression can be written as

$$|q|\int_0^{\pi}|\psi(\tau)|\mathrm{d}\tau \pm \left(\frac{1-a}{2}\right)\pi \leq \frac{1}{2m}\ln\frac{C}{\varepsilon} \qquad a \lessgtr 1 \tag{4.60}$$

where advantage has been taken of the fact that the integral of $|\psi(t)|$ over m periods is m times the integral over a single period.

It is interesting to note that the second term on the left hand side is always zero or positive so that a coarser condition is

$$|q|\int_0^{\pi}|\psi(\tau)|\mathrm{d}\tau \leq \frac{1}{2m}\ln\frac{C}{\varepsilon}. \tag{4.61}$$

By way of illustration it is interesting to apply Eq. (4.60) to the Meissner equation in which $\psi(t)$ is a unit rectangular waveform such as that shown in Fig. 3.1. For this waveform

$$\int_0^{\pi}|\psi(\tau)|\mathrm{d}\tau = \pi$$

so that Eq. (4.60) becomes

$$2|q| \pm (1-a) \leq \frac{1}{m\pi}\ln\frac{C}{\varepsilon} \qquad a \lessgtr 1. \tag{4.62}$$

This describes the relationship between a and q for the Meissner equation such that it is short term stable over m periods of the rectangular waveform, with respect to the parameters C and ε. It is shown as the cross-hatched region in Fig. 4.10 wherein it can be seen just how conservative a sufficiency criterion for short time stability can be. It is curious to note that it includes $a = 1$, $|q| > 0$

since this is a region of known classical instability. However if $m \to \infty$ the above inequality degenerates to $|q| = 0$, $a = 1$ which is classically stable point showing that the short time stability criterion is consistent, if relatively useless, in the limit.

For the Mathieu equation in which $\psi(t) = \cos 2t$, Eq. (4.60) becomes

$$\frac{4}{\pi}|q| \pm (1 - a) \leq \frac{1}{m\pi}\ln\frac{C}{\varepsilon}. \tag{4.63}$$

which is a more liberal condition as shown by the dotted lines in Fig. 4.10.

Condition 2. $M(t)$ in Eq. (4.59) has three principal minors; two are the same, viz. the diagonal elements of the matrix whilst the third is simply the determinant of $M(t)$. Consequently this sufficient condition for short time stability requires

$$\frac{1}{2T}\ln\frac{C}{\varepsilon} \geq 0$$

and

$$\left|\frac{1}{2T}\ln\frac{C}{\varepsilon}\right| \geq \left|\frac{a-1}{1} - q\psi(t)\right| \qquad 0 < t < T.$$

The first is always true since $T > 0$ and $C > \varepsilon$ by implication. As a result the modulus signs can be removed from the left hand side of the second expression. Thus we have

$$\begin{aligned}\frac{1}{2T}\ln\frac{C}{\varepsilon} &\geq \left|\frac{a-1}{2} - q\psi(t)\right| \\ &\geq \left|\frac{a-1}{2} - |q|\,|\psi(t)|\right|.\end{aligned}$$

Since $|\psi(t)|_{\max} = 1$ for a Hill equation the condition can be reduced to the more conservative form

$$\frac{1}{m\pi}\ln\frac{C}{\varepsilon} \geq \left|a - 1 - 2|q|\right|$$

where T has been replaced by $m\pi$ for observation over m periods of $\psi(t)$. This expression is essentially the same as Eq. (4.62)—i.e. condition 1 applied to the Meissner equation, and the region shown in Fig. 4.10 applies to condition 2, as modified, irrespective of the shape of $\psi(t)$.

Condition 3. From Eqs. (4.56) and (4.58) we have as this sufficient condition for the short time stability of a Hill equation

$$\int_0^T |1 - a + 2q\psi(t)|\,d\tau \leq \tfrac{1}{2}\ln\frac{C}{\varepsilon}$$

which can be recast as

$$\int_0^\pi |1 - a + 2q\psi(t)|\,d\tau \leq \frac{1}{2m}\ln\frac{C}{\varepsilon}$$

for $t = m\pi$. For the particular case of the Meissner equation, this last expression reduces to

$$|1 - a + 2q| + |1 - a - 2q| \leq \frac{1}{m\pi} \ln \frac{C}{\varepsilon}$$

which, when plotted on Fig. 4.10, will be seen to be more restrictive than the previous conditions, at least for this particular example.

References for Chapter 4

1. Willems, J. L.: Stability theory of dynamical systems. London: Wiley 1970
2. D'Angelo, H. D.: Linear time-varying systems: Analysis and synthesis. Boston: Allyn & Bacon 1970
3. Faddeeva, V. N.: Computational methods of linear algebra. New York: Dover 1959
4. McLachlan, N. W.: Theory and application of Mathieu functions. Clarendon: Oxford U. P. 1947. Reprinted by Dover, New York 1964
5. Whittaker, E. T.; Watson, G. N.: A course of modern analysis, 4th ed. London: Cambridge U. P. 1927
6. Meadows, H.: Solution of systems of linear ordinary differential equations with periodic coefficients. Bell Syst. Tech. J. 41 (1962) 1276–1294
7. Sato, C.: Stability conditions for resonant circuits with time-variable parameters. IRE Trans. Circuit Theory CT-9 (1962) 340–349
8. Keenan, R. K.: An investigation of some problems in periodically parametric systems. PhD Thesis, Monash University Melbourne, 1962
9. Arscott, F. M.: Periodic differential equations. Oxford: Pergamon 1964
10. Hochstadt, H.: A special Hill's equation with discontinuous coefficients. Am. Math. Monthly 70 (1963) 18–26
11. Hiller, J.; Keenan, R. K.: Stability of finite width sampled data systems. Int. J. Control 8 (1968) 1–22
12. Dorato, P.: Short-time stability in linear time-varying systems. IRE Int. Conv. Rec. PE 4 (1961) 83–87

Problems

4.1 Determine the range of a and q for which the solution to a Hill equation with a symmetric triangular coefficient with an initial condition of 1 will remain below 16 over 4 periods of the coefficient.

4.2 Are conditions for the short term stability for a Hill equation with a sawtooth coefficient dependent upon specifying whether the sawtooth has positive or negative slope?

4.3 A second order LC parallel network has $L = 100\ \mu\text{H}$ and $C = 20$ pF. Suppose the capacitor can be pumped periodically. Determine the four highest pumping frequencies that will induce instability in the network.

4.4 Verify theoretically the values of q at which the b_1, a_1 and b_2 boundaries intersect the q axis of the stability diagram for the Meissner equation with $\tau = \pi/2$.

4.5 Determine the range of q over which the Hill equation with an impulsive coefficient as defined in Eq. (4.10) has a stable solution, if $a = 0$.

4.6 A child on a playground swing generally pumps once in every oscillation. Does he swing in a $2P$ mode or a $2N$ mode?

4.7 A particular high order Hill equation is

$$\overset{v.}{x} + (a - 2q\psi(t))x = 0, \ \psi(t) = \psi(t + \theta)$$

where $\overset{v.}{x}$ is the vth derivative of x with respect to t. Demonstrate that mode boundary intersections in an a, q mode diagram, occur when $a = n^v$, where n is an integer. (This has been known as Keenan's conjecture and was treated by Richards and Miller in SIAM J. Appl. Math. 25 (1973) 72–82. It can be verified readily using Eq. (4.42)).

4.8 For a general second order Hill equation verify that the only mode of solution possible for $a < 0$ and q small is a $2P$ mode.

4.9 How would you expect the mode diagram of Fig. 4.9 to be modified if the static system has two real poles and one complex conjugate pair?

4.10 Figure 4.4 shows that the equation of Eq. (4.10) has stable solutions for $a = 0$ only if $-\frac{2}{\pi} \leq q < 0.$

Verify this range (to a first approximation) using a 3×3 truncated Hill determinant. This will require finding the Fourier coefficients of a periodic sequence of impulses.

4.11 Using the trace of the state transition matrix demonstrate that the Meissner equation has a periodic solution at the points
(i) $a = 4, q = 0$
(ii) $a = 0, q = 0.707$
(iii) $a = 2, q = -1$ (approximate)

4.12 The periodic coefficient in a particular Hill equation is -1 over a complete period except for a segment, equal to 20% of the period, where it is $+1$. Demonstrate that the equation's stability is not affected by the position at which the positive portion appears in the period.

Chapter 5

A Modelling Technique for Hill Equations

Surprisingly few Hill equations are analytically tractable as can be assessed from the relatively small number treated in Chap. 3. In fact the most commonly encountered Hill equation—viz. the Mathieu equation—belongs in that class. In practice therefore physical descriptions are sometimes compromised to render the equation solvable, or else only approximate solutions to the equations are sought. Clearly, in each case, essential information may be lost, or misleading results may be obtained, especially in the vicinity of points of marginal stability (the stability boundaries of Chap. 4) where system sensitivity to features of the periodic coefficient can be very high. Numerical integration techniques are also often used with difficult Hill equations; however these tend to be time consuming and this is a particular disadvantage in cases where many solutions might be required. Moreover accurate assessment of stability cannot be made from numerical solutions. Certainly alternative methods of treatment are required that will yield accurate solutions with a minimum of (computational) effort and which will also permit the stability of Hill equations to be readily assessed.

The techniques developed in this chapter are based upon modelling or replacing a particular intractable Hill equation by a counterpart which exhibits very similar solutions and stability properties and yet which can be handled exactly mathematically. These modelling procedures are developed by demonstrating, first, that the behaviour of a Hill equation, especially for the ranges of its coefficients encountered in most practical situations, is determined principally by the lower order harmonics in its periodic coefficients with higher harmonics having a progressively decreasing influence as they become of higher order. Secondly the approach rests upon being able to identify classes of periodic coefficient in which harmonic content can be readily adjusted and for which the corresponding Hill equation is tractable.

5.1 Convergence of the Hill Determinant and Significance of the Harmonics of the Periodic Coefficients

5.1.1 Second Order Systems

Equation (2.21) shows that a fundamental solution to a (homogeneous) Hill equation is of the form

$$x(t) = \exp(\mu t)\Xi(t)$$

where the characteristic exponent μ is determined using a doubly infinite determinant, $\Delta(\mu)$, as seen in Sect. 4.3.6. For lossless second order systems the corresponding Hill determinantal equation can be used to derive, as an equation for deducing μ, the expression

$$\cosh \mu\pi = 1 - 2\Delta(0) \sin^2 \{\pi/2\sqrt{G_0}\}, \; G_0 = a - 2q\Psi_0, \qquad (5.1a)^1$$

where Ψ_0 is the average value of the periodic coefficient and $\Delta(0)$ is the value of $\Delta(\mu)$ at $\mu = 0$, as defined in Eq. (4.20).

The denominator terms in $\Delta(\mu)$ are of the form

$$\rho_r = G_0 - (2r - j\mu)^2, \quad j = \sqrt{-1} \qquad (5.1b)$$

giving as the denominators in $\Delta(0)$

$$\rho_r' = G_0 - (2r)^2 \qquad (5.1c)$$

where r is the row index integer in the determinants. As a result of the squared terms in Eqs. (5.1) both $\Delta(\mu)$ and $\Delta(0)$ are convergent about their central elements; this is reinforced usually by the convergence of the sequence of Fourier coefficients G_i, shown in Eqs. (4.13) and (4.16)[2]. Therefore the contributions of determinantal elements to the values of the determinants diminish as the element lies further from the principal diagonal. Consequently, owing to the structures of the determinants, contributions from the harmonics of the periodic coefficient will decrease as the harmonics become of higher order. Ultimately harmonics beyond a particular order can be taken to have a negligible effect on the values of $\Delta(\mu)$ and $\Delta(0)$, and thus on the value of the characteristic exponent μ and the actual solution to the Hill equation. In some applications this point can be determined and the infinite determinants truncated to give a workable finite approximation. For example, Hill [1], in his celebrated astronomical study, used 3 × 3 truncated Hill determinants with little error.

As an illustration of the decreasing importance of the harmonics of the periodic coefficient, Table 5.1 shows values of the characteristic exponent calculated at a number of values of a and q for the case of a sinusoidal coefficient with various additional amounts of third, fifth and seventh harmonics. The values of a and q have been chosen as representative of those encountered in practice. For example $a = 1.0$, $q = 0.1$ is typical of degenerate parametric amplification, $a = 0.23699$, $q = 0.70600$ falls very near a stability boundary and $a = 1.0$, $q = 1.5$ is a case where the time-varying portion of the coefficient dominates over the constant term. In each case it is evident that the influence of the harmonics decreases as they become of higher order.

[1] Lossy second order systems can also be treated using this expression, after having been transformed according to Eq. (2.4).

[2] In (5.1c) it is apparent that $\Delta(0)$ has poles at $G_0 = (2r)^2$. These do not affect the value of μ in any discontinuous fashion since they are compensated by the zeros of the sinusoidal term in (5.1a).

Table 5.1. Effect of harmonic distortion on the characteristic exponent. (Proc. IEEE 65 (1977) 1549–1557)

Harmonic content			
a	1.0	0.23699	1.0
q	0.1	0.70600	1.5
0	0.04994 + j0.0	0.00227 + j0.0	0.62184 + j0.0
3rd harmonic			
5%	0.04994 + j0.0	0.01570 + j0.0	0.62323 + j0.0
15%	0.04994 + j0.0	0.03116 + j0.0	0.62584 + j0.0
25%	0.04995 + j0.0	0.04489 + j0.0	0.62803 + j0.0
5th harmonic			
5%	0.04994 + j0.0	0.00432 + j0.0	0.62182 + j0.0
15%	0.04994 + j0.0	0.01122 + j0.0	0.62168 + j0.0
25%	0.04994 + j0.0	0.01843 + j0.0	0.62138 + j0.0
7th harmonic			
5%	0.04994 + j0.0	0.00344 + j0.0	0.62183 + j0.0
15%	0.04994 + j0.0	0.00808 + j0.0	0.62174 + j0.0
25%	0.04994 + j0.0	0.01313 + j0.0	0.62158 + j0.0

Two further observations can be made from Table 5.1. First, it will be observed that the characteristic exponent is most sensitive to harmonics in the vicinity of marginal stability. The reason for this can be seen by considering the derivative (from Eq. (5.1a))

$$\frac{d\mu}{d\Psi_i} = \frac{d\mu}{d\Delta(0)} \cdot \frac{d\Delta(0)}{d\Psi_i}$$

$$= \frac{-\pi G_0}{2 \sinh \mu\pi} \cdot \left\{\frac{\sin x}{x}\right\}^2 \cdot \frac{d\Delta(0)}{d\Psi_i}, \quad x = \frac{\pi}{2}\sqrt{G_0}.$$

Thus when μ is in the vicinity of $0 \cdot 0 + jm$, m integral it is most sensitive to changes in $\Delta(0)$ and thus to changes or errors in the Ψ_i. The degree of this sensitivity requires determination of $d\Delta(0)/d\Psi_i$ which is difficult in general terms. For particular applications however this factor might be determined numerically, if necessary, simply by evaluating $\Delta(0)$ for a range of Ψ_i at the particular values of a and q of interest.

The second remark that can be made concerns the influence of the value of q on the sensitivity of $\Delta(0)$ to the harmonics Ψ_i. The magnitude of q determines the strengths of the off-diagonal elements in the Hill determinants with respect to unity (the magnitudes of the diagonal terms). For small q the sensitivity of $\Delta(0)$ to all harmonics is low, whereas for large q many $d\Delta(0)/d\Psi_i$ may be significant, implying that the range of significant harmonics will be extended, as can be observed in Table 5.1. For q small the influence of seventh harmonic is less than that of fifth whereas for $q = 1.5$, seventh harmonic is only slightly less important than fifth.

A final comment relates to the importance of a. For this purpose let $\Psi_0 = 0$ so that $G_0 = a$. For small a (<2) the denominator terms form a monotonically increasing sequence so that sensitivity to harmonics decreases monotonically as the harmonics become of higher order, in accord with earlier comments. However for larger a this uniform behaviour is disrupted, leading to an increased dependence on a lower range of harmonics before sensitivity then decreases monotonically and convergence of the determinant commences. The particular harmonic beyond which strong convergence of the determinant occurs can be ascertained by studying the divergence of $a - (2r)^2$, with r. For non-zero Ψ_0 this discussion can be modified appropriately.

5.1.2 General Order Sytems

When treating periodically time varying systems of general order a doubly infinite Hill determinant can also be derived as illustrated in Sect. 4.4. The denominator terms in this are of the form

$$\rho_r = \sum_{k=0}^{v} \alpha_k(\mu + jr\omega_p)^k \tag{5.2}$$

in which the order of the system is v and r is the row index in the determinant. ω_p is the fundamental radian frequency of the time variation and α_k is the constant portion of the time-varying coefficient $a_k(t)$ accociated with the kth derivative. As in the particular case of second order systems discussed in Sect. 5.1.1 the magnitude of ρ_r, as a function of r, determines convergence of the determinant. Convergence will not occur until r is sufficiently large that ρ_r is clear of all its roots and is diverging to infinity. This particular value of r will have to be identified for each particular system. This will then specify the upper range of harmonics of the time variation to which the system will be insensitive. It is of interest to observe that Eq. (5.2) is of the form of the characteristic polynominal of the static (i.e. constant coefficient) system so that the value of r could be deduced from an observation of the frequency at which the characteristic equation diverges—i.e. from an examination of the frequency response of the static system.

5.2 A Modelling Philosophy for Intractable Hill Equations

Since the behaviour of a periodic differential equation is dependent primarily upon the lower frequency content of its periodic coefficients and decreases in its dependence upon harmonics as they become of higher order, it is possible to have two Hill equations with differing coefficients but very similar solutions. Indeed only the lower frequency spectra of their periodic coefficients need to be the same to ensure this. This suggests then that intractable Hill equations could be modelled by tractable counterparts provided the lower frequency spectra of their coefficients are very similar. For such an approach to be workable it is important that there exist a class of Hill equation which, in addition to being tractable, has periodic coefficients that can be manipulated easily to produce different harmonic compositions. Perhaps the most versatile class of this nature is that of Hill equations with staircase coefficients, as discussed in Sect. 3.5. A further class is that of Hill equations with piecewise linear coefficients, although it is generally more complex to handle algebraically than the former (see Sect. 5.4).

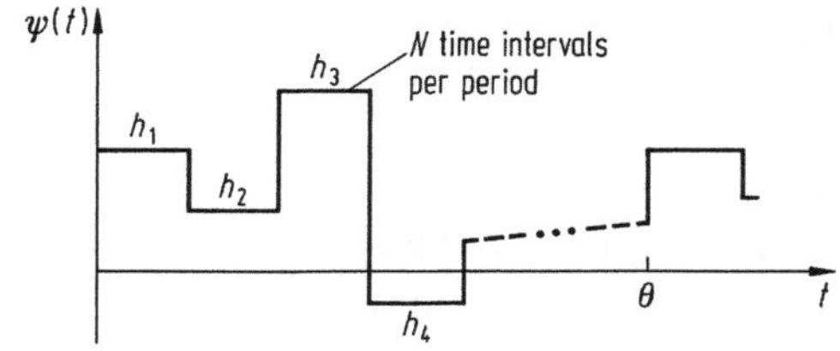

Fig. 5.1 Staircase waveform with equal time steps

Fig. 5.2 Unit periodic sampling function

5.3 The Frequency Spectrum of a Periodic Staircase Coefficient

In order to use a Hill equation with a piecewise constant or staircase coefficient as a successful model it is necessary to know how to order the steps in the coefficient to produce a desired frequency spectrum. Using the development of Franks and Sandberg [2] it is fortunately very easy to specify the characteristics of such a coefficient from a harmonic component prescription.

Consider the general periodic staircase waveform illustrated in Fig. 5.1 in which the time intervals are all equal. For a period of θ as shown, and N time segments, each segment will be of length θ/N. If the step heights are designated $h_1, h_2, \ldots, h_N$ from the axis then the function can be expressed

$$\psi(t) = \sum_{n=1}^{N} h_n W\{t - (n-1)\theta/N\} \tag{5.3}$$

where $W(t)$ is the unit periodic sampling function shown in Fig. 5.2 and defined by

$$W(t) = 1 \text{ for } k\theta < t < k\theta + \theta/N, \quad k\text{-integral}$$
$$= 0 \text{ elsewhere.}$$

The complex frequency spectrum of $\psi(t)$ is given by

$$\Psi_m = \Psi(m\omega_0) = \frac{1}{\theta}\int_0^\theta \psi(t)\mathrm{e}^{-\mathrm{j}2\pi mt/\theta}\mathrm{d}t, \quad m \text{ integral}$$

with $\omega_0 = 2\pi/\theta$. This expression becomes, using Eq. (5.3),

$$\Psi_m = \frac{1}{\theta}\int_0^\theta \sum_{n=1}^{N} h_n W\{t - (n-1)\theta/N\}\mathrm{e}^{-\mathrm{j}2\pi mt/\theta}\mathrm{d}t$$
$$= \frac{1}{\theta}\sum_{n=1}^{N} h_n \int_0^\theta W\{t - (n-1)\theta/N\}\mathrm{e}^{-\mathrm{j}2\pi mt/\theta}\mathrm{d}t$$
$$= \frac{1}{\theta}\sum_{n=1}^{N} h_n \int_{(n-1)\theta/N}^{n\theta/N} \mathrm{e}^{-\mathrm{j}2\pi mt/\theta}\mathrm{d}t$$

which, upon evaluation, can be written as

$$\Psi_m = \sum_{n=1}^{N} h_n \frac{e^{jm\pi/N} \sin m\pi/N}{m\pi} \cdot e^{-j2\pi mn/N} \cdot h_n. \tag{5.4}$$

The Ψ_m obey a particularly useful and simple recurrence relationship. Consider, with k integral,

$$\Psi_{m+kN} = \sum_{n=1}^{N} \frac{e^{j(m+kN)\pi/N} \sin(m+kN)\pi/N \, e^{-j2\pi(m+kN)n/N}}{(m+kN)\pi} \cdot h_n$$

i.e.

$$\Psi_{m+kN} = \frac{m}{m+kN} \Psi_m. \tag{5.5}$$

Therefore once the first $N - 1$ coefficients have been determined all other Ψ_m can be evaluated. Moreover the higher harmonics are decreasing approximately as $1/k$, a feature which should be compared to the same rate of decrease known for the coefficients of a square waveform.

Whilst Eq. (5.4) allows a particular staircase coefficient to be analysed for its frequency spectrum, it is really the inverse problem that is of interest to the present modelling procedure—i.e. given a prescribed coefficient frequency spectrum that is significant in the response of a particular parametric system, synthesise a corresponding piecewise-constant coefficient. As with the spectral analysis above, this is a moderately straightforward procedure and has been carried out by Franks and Sandberg [2] to yield

$$h_n = \sum_{\substack{m \\ |m|<N/2}} \frac{e^{-jm\pi/N} \cdot m\pi/N}{\sin m\pi/N} \cdot e^{j2\pi mn/N} \cdot \Psi_m \tag{5.6}$$

with $\Psi_{N/2} = 0$ if N is even. Equation (5.6) is central to staircase models for the coefficients in Hill equations. As an illustration of its application consider the generation of a staircase approximation to the sinusoid in the Mathieu equation. For a sinusoid $\Psi_{\pm 1} = \mp j/2$ and $\Psi_{\pm i} = 0$, $i = 2 \ldots N/2$. This leads to step heights given by

$$h_n = \sin(2n-1)\pi/N \cdot \frac{\pi/N}{\sin \pi/N}, \quad n = 1, \ldots, N. \tag{5.7}$$

Equation (5.7) is used extensively in Chap. 6 when dealing with specific models for the Mathieu equation and their relative merits.

Pipes [3] appears to have been the first to demonstrate the value of piecewise constant Hill equation models in treating such intractable forms as the Mathieu equation. In the absence of any alternative guideline however he suggested choosing step heights as the average value, over the corresponding time interval, of the time-varying coefficient to be emulated. Thus if $\psi(t)$ is the actual periodic coefficient then over the nth interval the model step height would be

$$h_n = \frac{N}{\theta} \int_{(n-1)\theta/N}^{n\theta/N} \psi(t) \mathrm{d}t. \tag{5.8}$$

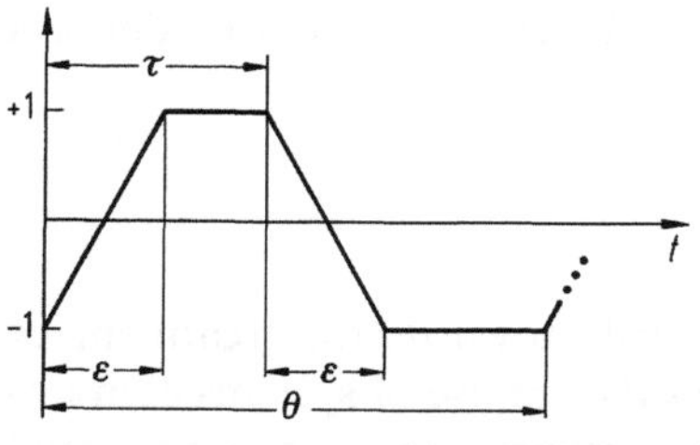

Fig. 5.3. Unit trapezoidal waveform with equal rise and decay times

Although equal time intervals would generally be chosen non-uniform segments are clearly also acceptable in this case.

Whilst the adoption of Eq. (5.8) might be a useful means for devising a set of h_n when the frequency spectrum of $\psi(t)$ is not known or cannot be readily determined, it does not lead, in general, to as good a model as that based upon Eq. (5.6). This is pursued in detail for the Mathieu equation in Chap. 6.

5.4 Piecewise Linear Models

5.4.1 General Comments

Owing to the added complexity in programming introduced by the need to generate and evaluate fractional order Bessel functions, piecewise linear models are perhaps at first impression less attractive than those based upon staircase waveforms. However for a given number of time segments a piecewise linear model can give much better results, as demonstrated in Sect. 6.3.4, in which a trapezoidal waveform (requiring $N = 4$ time steps) is shown to lead to a considerably better model for the Mathieu equation than a staircase model with even $N = 10$ time intervals. This is because the trapezoid has more parameters that can be adjusted for a given number of time steps than a staircase function, in order to tailor its frequency spectrum. Indeed the trapezoidal waveform has available N addition parameters in the gradients of its segments, as have all piecewise linear functions. Generalised expressions for the spectral components of piecewise linear waveforms can be derived in a manner similar to that used in Sect. 5.3 when dealing with staircase coefficients.

5.4.2 Trapezoidal Models

For the trapezoidal waveform illustrated in Fig. 5.3 it is readily shown that the mth complex Fourier coefficient is given by

$$\Psi_m = -\{1 - \mathrm{e}^{-\mathrm{j}2\pi m\delta}\}\{1 - \mathrm{e}^{-\mathrm{j}2\pi m\eta}\}/2m^2\pi^2\eta \tag{5.9}$$

where $\delta = \tau/\theta$ and $\eta = \varepsilon/\theta$.

Inspection of Eq. (5.9) reveals that the spectrum of the trapezoid can be modified over a considerable range by choosing selectively values for δ or η. For example if either δ or η are chosen as $1/k$, where k is an integer then all harmonics of order $m = kn$, where n is an integer, will vanish. As an illustration $\delta = \frac{1}{2}$ will remove all even ordered harmonics, as is well known. Furthermore if δ is chosen as $1/k_1$ and η chosen as $1/k_2$ then all harmonics of orders $m = k_1 n$ and $m = k_2 n$ will vanish. In Sect. 6.3.2 δ is chosen as $\frac{1}{2}$ whereas η is chosen as 1/3 in a model of a sinusoid.

5.5 Forced Response Modelling

The preceding sections have been directed primarily towards the treatment of homogeneous Hill equations and thus to the *natural* response of systems containing a periodically variable parameter. When such a system is driven by an external input, its complete response is given from Eq. (2.13), as

$$\boldsymbol{x}(t) = x_N(t) + x_F(t) = \phi(t, 0)\boldsymbol{x}(0) + \int_0^t \phi(t, \tau)\boldsymbol{f}(\tau)\mathrm{d}\tau \tag{5.10}$$

where $\boldsymbol{f}(\tau)$ is a vector which accounts for the input. The first (left hand) term in the solution is the natural response which has been the subject of the discussion above, in algebraic rather than matrix form. The second term in Eq. (5.10) is the forced response.

The state transition matrix $\phi(t, 0)$ contains all the information relevant to the properties of a particular system including the magnitudes of the parameters and the nature of the time variation. The material of the foregoing sections can therefore be viewed as means for obtaining an expression for $\phi(t, 0)$ in the face of intractabilities. Since that transition matrix appears also in the forced response term it is clear that the same modelling procedures can be employed to simulate the forced response of an intractable periodic system. There then arises the question of how easily the integral in Eq. (5.10) can be evaluated, and this will depend upon the particular form of $\boldsymbol{f}(t)$, which clearly precludes general comment.

5.6 Stability Diagram and Characteristic Exponent Modelling

Determination of the characteristic exponent for a given set of parameters, or even a complete stability diagram, is straightforward and follows the same modelling/procedure as for the natural solutions. Again an equation is chosen such that its periodic coefficient has a lower frequency spectrum which resembles that of the coefficient of the intractable equation of interest. For this substitute equation the discrete transistion matrix $\phi(\theta, 0)$ is derived and its eigenvalues are noted to determine stability according to the guidelines laid down in Sect. 4.2. This technique has been used by Lee [4] to determine $\phi(\theta, 0)$ where the step heights were chosen according to Eq. (5.8).

Characteristic exponents can be determined from Eq. (2.22) once the discrete transition matrix and thus its eigenvalues are known for a given set of parameters. Alternatively Eq. (2.27b) can be used for second order systems.

5.7 Models for Nonlinear Hill Equations

Parametric systems with nonlinearities which lead to a Hill equation description of the form

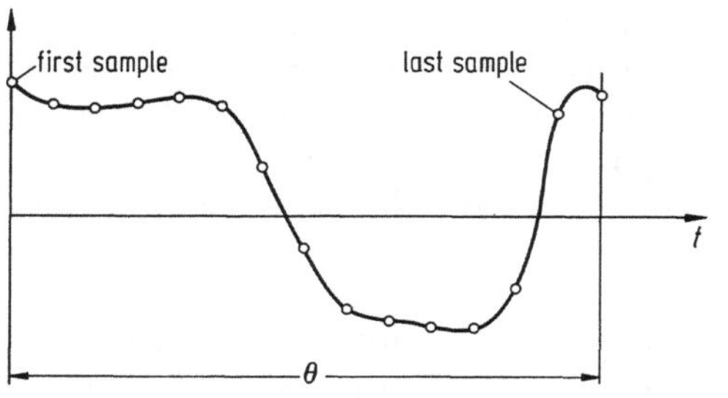

Fig. 5.4. Sampling a periodic function for discrete spectral analysis

$$\ddot{x} + g(t)(x + b'x^2 + c'x^3 + \cdots) = 0, \quad g(t) = g(t + \pi) \qquad (5.11)$$

are encountered in the detailed analyses of parametric amplification and dynamic focussing using quadrupole lenses (see Chap. 7). Clearly, owing to the combined time-variable coefficients and nonlinearities, these equations are always treated using techniques for numerical integration which demand considerable computational time if a high degree of accuracy is required. Modelling $g(t)$ can be of benefit however in these cases, as well as for their linear counterparts. Replacement of $g(t)$ by a piecewise constant model resolves Eq. (5.11) into a set of time contiguous, ordinary nonlinear equations of the form

$$\ddot{x} + ax + bx^2 + cx^3 + \cdots = 0$$

which may be treated by a number of standard techniques that are found discussed in Cunningham [5] and Minorsky [6].

5.8 A Note on Discrete Spectral Analysis

In some situations, particularly experimental, it may be difficult to express the periodic coefficients in a Hill equation in mathematical terms, thus apparently precluding spectral analysis if a modelling-oriented solution is to be sought. (Indeed in such situations it is difficult to foresee any other method to be of value.) Moreover in other situations even though explicit mathematical expressions may be available for the coefficients, the corresponding integration of the Fourier series coefficient formulae may be difficult. In these cases numerical methods for spectral analysis can be used to particular advantage, in which samples of the coefficients are taken over one complete period and fed to a computer algorithm for production of the spectrum.

Most program libraries of large computer centres contain routines for numerical or so-called discrete spectral analysis. Usually these are based upon the fast Fourier transform algorithm, which is a particularly fast and efficient procedure. However, irrespective of the technique employed, discrete spectral analysis can have its pitfalls if not used correctly. Fortunately, for the analysis of periodic waveforms these are minimised, the only one of consequence being *aliasing*—i.e. errors introduced into the spectral components by not taking enough samples of the waveform in a given time.

When using discrete Fourier transform routines to determine the frequency spectra of periodic coefficients, the waveforms should be sampled over one, or possibly an integral number, of periods, finishing one sample interval short of the point on the waveform at which sampling was commenced, as depicted in Fig. 5.4. This is necessary since the algorithms consider the sequence of samples fed to them as one period of an infinite periodic series, of period defined by the time range of the samples.

The number of samples should be chosen in the first instance to ensure that sufficient spectral components have been provided. If θ is the period of sampling and N the number of samples taken then the highest frequency component provided by discrete spectral analysis will be

$$F_{\text{upper}} = N/2\theta.$$

To avoid aliasing entirely it is necessary to ensure (*a priori*) that the waveform possess no frequency components beyond this. For some functions this is easy, as in the case of a sinusoid or a finite sum of sinusoids. However in general it may be impossible to choose F_{upper} sufficiently high to encompass all frequencies. For example a square waveform has all frequency components, of odd order, to infinity. In these cases it may be necessary to determine empirically a reasonable F_{upper} beyond which all remaining frequency components are of negligible magnitude. With a square wave, for instance, the components decrease according to $1/k$ where k is their order, so that the 21st component has an amplitude less than 5% of the first. This may be considered sufficiently small to neglect all higher terms. In general however it may be a trial and error process to ascertain that aliasing is negligible. Perhaps a useful guideline is to ensure that the component at F_{upper} and the preceding few should be of negligible amplitude. If this is the case aliasing should have been minimised unless the waveform contains some unusually large (unsuspected) higher harmonics. This is seldom the case, especially in a practical situation where system time constants often impose a low pass filter constraint on signals.

A very detailed and readable treatment of discrete spectral analysis for all waveform types, and of the fast Fourier transform algorithm, is found in the book by Brigham [7].

5.9 Concluding Remarks

The concept of treating an intractable Hill equation by a readily handled counterpart appears to have been used first by van der Pol and Strutt [8] in an analysis of the properties of the Mathieu equation. For this they replaced the sinusoid by a square waveform and subsequently studied properties of the ensuing Meissner equation. Pipes [3] revived the method in 1955, emphasising the ease with which certain (tractable) Hill equations can be treated using matrix techniques. Even though he emphasised the value of these in modelling difficult equations, he did not consider the significance to the solution of the harmonic composition of the coefficients and the considerably simpler models which can result. Since then the technique seems to have had its main application to the Mathieu equation although there are many experimental situations, particularly, where the method can be of value on more complex waveshapes.

The emphasis of the technique, as developed in this chapter, has been on synthesising models via the frequency domain. However there may be many situations in practice where, by inspection, a time domain fitting of constant, linear and even exponential segments to a particular wave-shape might easily be performed in a pursuit of a tractable model. For example an experimentally generated rectangular waveform will have rising exponential like edges and may even exhibit overshoot and droop all of which could be accommodated by a combination of steps and ramps.

References for Chapter 5

1. Hill, G. W.: On the part of the moon's motion which is a function of the mean motions of the sun and the moon. Acta Math. 8 (1886) 1–36
2. Franks, L. E.; Sandberg, I. W.: An alternative approach to the realization of network transfer functions: The N-path filter. Bell Syst. Tech. J. 89 (1960) 1321–1350
3. Pipes, L. A.: Matrix solution of equations of the Mathieu—Hill type. J. Appl. Phys. 24 (1953) 902–910
4. Lee, I.: On the theory of linear dynamic systems with periodic parameters. Inf. Control 6 (1963) 265–275
5. Cunningham, W. J.: Nonlinear equations. New York: McGraw—Hill 1958
6. Minorsky, W.: Nonlinear oscillations. New York: Van Nostrand 1962
7. Brigham, E. O.: The fast Fourier transform. Englewood Cliffs: Prentice Hall 1974
8. van der Pol. B.; Strutt, M. J. O.: On the stability of solutions of Mathieu's equation. Philos. Mag. 5 (1928) 18–39

Problems

5.1 The characteristic exponent for the Hill equation

$$\ddot{x} + (a - 2q\psi(t))x = 0, \quad \psi(t) = \psi(t + \theta)$$

can be obtained from

$$\cosh 2\pi\mu/\omega_p = 1 - 2\Delta(0)\sin^2\{\pi\sqrt{G_0}/\omega_p\} \quad \omega_p = 2\pi/0$$

where $G_0 = a - 2q\Psi_0$ and $\Delta(0)$ is defined by Eqs. (4.18) and (4.20) with $\mu = 0$. Show that the poles of $\Delta(0)$ at $G_0 = -(\omega_p r)^2$, where r is the row index, are removed by zeros of $\sin^2\{\pi\sqrt{G_0}/\omega_p\}$.

5.2 Use Eq. (5.6) to determine a six step approximation to

(i) a sinusoid
(ii) a positive slope sawtooth
(iii) a negative slope sawtooth

5.3 Evaluate Eq. (5.8) for the case of a sinusoid and show that Eqs. (5.8) and (5.7) lead to the same results only if N, the number of steps, goes to infinity.

5.4 Use Eq. (5.1a) to determine the value of the characteristic exponent for the Mathieu and Meissner equations at the points
(i) $a = 0, q = 0.8$
(ii) $a = 1, q = 0.1$.
To do this, approximate $\Delta(0)$ by a finite 3×3 determinant.

5.5 Verify that the characteristic exponent for a Hill equation, in the vininity of the origin in the a, q stability diagram, can be obtained from

$$\cosh \mu\pi = 1 - 2\{G_0 + 2q^2\Psi_1\Psi_1^*\}\{(\tfrac{\pi}{2})^2 - \tfrac{1}{3}(\tfrac{\pi}{2})^4 G_0\}$$

where $G_0 = a - 2q\Psi_0$ and Ψ_0 and Ψ_1 are the first two complex Fourier coefficients of the periodic coefficient.

Show that the a_0 stability boundary for $\Psi_0 = 0$ near the origin, is parabolic and approximated by

$$a + 2q^2\Psi_1\Psi_1^* = 0.$$

5.6 Use the expression for μ in Prob. 5.5 to demonstrate, at least near the origin, that a stability diagram will be asymmetric with q if $\Psi_0 \neq 0$.

5.7(a) It is proposed to use the pulse wavegorm shown below as a substitute for the sinusoid in the Mathieu equation. Determine its complex Fourier series and hence choose A so that it has a fundamental magnitude equal to that of a unity amplitude sinusoid.

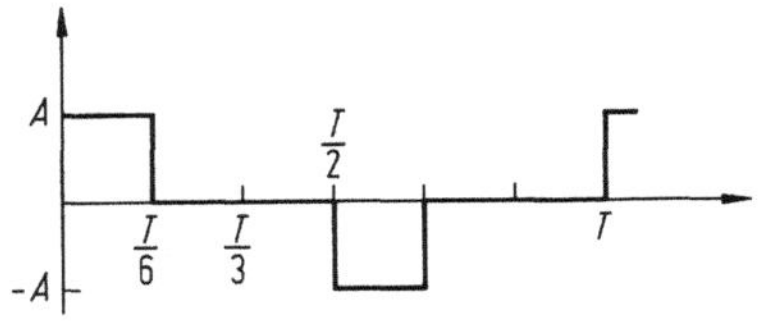

(b) Determine the discrete transition matrix for the above pulse waveform (with the value of A you have chosen) and derive an expression by which stability of the Mathieu equation can be assessed. From this calculate values for the characteristic exponent at
(i) $a = 0, q = 0.8$
(ii) $a = 1, q = 0.1$
and compare the results to those from Prob. 5.4.

5.8 Determine suitable two step approximations for negative and positive sawtooth waveforms and use the approximations to demonstrate that the stability of a Hill equation with a sawtooth waveform is independent of whether the waveform has positive or negative slope; (cf. Prob. 4.2).

Chapter 6

The Mathieu Equation

The Mathieu equation in its standard form

$$\ddot{x} + (a - 2q\cos 2t)x = 0 \tag{6.1}$$

is the most widely known and, in the past, most extensively treated Hill equation. In many ways this is curious since the equation eludes solution in a usable closed form; yet many investigators have sought to describe experiments in terms of a Mathieu equation, most probably only because it contains a simple sinusoid as its periodic coefficient. By association with Fourier series it may have been assumed that once solutions to the Mathieu equation had been determined, solutions to Hill equations in general would follow. Indeed, in many ways the opposite is true in the context of the methods presented in Chap. 5.

This chapter describes various means for treating the Mathieu equation. Commencing with a brief review of classical techniques it is shown that the usefulness of the exact Mathieu equation and its solutions is limited. It is shown, instead, that by modelling the sinusoidal coefficient in the manner of the techniques presented in Chap. 5, equations very similar to the Mathieu equation can be deduced and solved, and consequently are of more practical value. Thus rather than seeking approximate solutions to the exact Mathieu equation, exact solutions to an approximate Mathieu equation will be exploited. The latter approach has much to recommend it computationally.

6.1 Classical Methods for Analysis and Their Limitations

The most comprehensive treatment of classical methods for the analysis of the Mathieu equation has been given by McLachlan [1]. His book covers both the Hill determinant procedures outlined in Chap. 4, and their outcomes, and also presents solutions based upon assumed series forms. Some aspects of the latter are pursued in the following section to illustrate the method.

6.1.1 Periodic Solutions

The starting point for the series expansion approach to obtaining solutions to the Mathieu equation is to observe that these solutions must reduce to either $\cos mt$ or $\sin mt$, $m^2 = a$, for $q \to 0$.—i.e. the equation and its solutions reduce to those of simple harmonic motion. As q is increased from zero the basic solution must be modified to account for the degree of periodic coefficient that has been introduced.

At $a = 1$ and $q = 0$ the basis solutions are $\cos t$ and $\sin t$. Concentrating, for the present, upon the even solution as a starting point, a general solution for q non zero can be expressed therefore as

$$x(t) = \cos t + qC_1(t) + q^2C_2(t) + \cdots \tag{6.2}$$

where the functions $C_i(t)$ are to be determined. The first task in the classical theory generally is to determine solutions that are periodic for all q. To maintain periodicity as q increases above zero the value of a may need continual modification with q. To permit this, a is also expressed as an infinite series in q,

$$a = 1 + \alpha_1 q + \alpha_2 q^2 + \cdots. \tag{6.3}$$

The constants α_i and the functions $C_i(t)$ can be determined by substituting Eqs. (6.2) and (6.3) into Eq. (6.1), equating coefficients of powers of q to zero and by removing non-periodic terms. Upon so doing, it is found that the first periodic solution of the Mathieu equation is

$$\begin{aligned} x(t) = \cos t - \frac{1}{8}q\cos 3t + \frac{1}{64}q^2(-\cos 3t + \frac{1}{3}\cos 5t) \\ - \frac{1}{512}q^3(\frac{1}{3}\cos 3t - \frac{4}{9}\cos 5t + \frac{1}{18}\cos 7t) + \cdots \end{aligned} \tag{6.4}$$

with a constrained to

$$a = 1 + q - \frac{1}{8}q^2 - \frac{1}{64}q^3 - \frac{1}{1536}q^4 + \cdots. \tag{6.5}$$

The solution described in Eq. (6.4) is called a *Mathieu Function* (of the first kind) and is denoted $ce_1(t, q)$. For a given value of q, the value of a generated by Eq. (6.5) at which the corresponding Mathieu function exists, is called a *characteristic number*.

Commencing with the odd solution $\sin t$ for $a = 1$, $q = 0$, another general periodic solution to the Mathieu equation is

$$\begin{aligned} se_1(t, q) = \sin t - \frac{1}{8}q\sin 3t + \frac{1}{64}q^2\left(\sin 3t + \frac{1}{3}\sin 5t\right) \\ - \frac{1}{512}q^3\left(\frac{1}{3}\sin 3t + \frac{4}{9}\sin 5t + \frac{1}{18}\sin 7t\right) + \cdots \end{aligned} \tag{6.6}$$

provided

$$a = 1 - q - \frac{1}{8}q^2 + \frac{1}{64}q^3 - \frac{1}{1536}q^4 + \cdots. \tag{6.7}$$

Inspection of Eqs. (6.5) and (6.7) reveals that the solutions given in Eqs. (6.4) and (6.6) do not coexist except when $q = 0$. For $q \neq 0$, the values of a for which the periodic solutions exist are quite different. The second linearly independent solutions for non-zero q are shown by McLachlan [1] to be non periodic and, in fact, of the t-multiplied variety—i.e. they are marginally unstable. Indeed it can be demonstrated that periodic solutions of the types above exist only on stability boundaries, with the a, q relationships in Eqs. (6.5) and (6.7) therefore being polynomial expressions for those boundaries. In particular Eq. (6.5) describes the a_1 boundary in the Mathieu equation stability diagram and Eq. (6.7) describes the b_1 boundary (cf. boundaries for the Meissner equation in Fig. 4.2).

In a manner analogous to that above a complete hierarchy of periodic solutions of odd and even type, along with their associated characteristic numbers, can be established. The first few members in this hierarchy are

$$ce_0(t, q) = 1 - \frac{1}{2}q\cos 2t + \frac{1}{32}q^2\cos 4t - \cdots$$

$$ce_1(t, q) = \cos t - \frac{1}{8}q\cos 3t + \frac{1}{64}q^2\left(-\cos 3t + \frac{1}{3}\cos 5t\right) - \cdots$$

$$se_1(t, q) = \sin t - \frac{1}{8}q\sin 3t + \frac{1}{64}q^2\left(\sin 3t + \frac{1}{3}\sin 5t\right) - \cdots$$

$$ce_2(t, q) = \cos 2t - \frac{1}{8}q\left(\frac{2}{3}\cos 4t - 2\right) + \frac{1}{384}q^2\cos 6t - \cdots$$

$$se_2(t, q) = \sin 2t - \frac{1}{12}q\sin 4t + \frac{1}{384}q^2\sin 6t - \cdots$$

and so on, with the corresponding characteristic numbers

$$a_0\colon a = -\frac{1}{2}q^2 + \frac{7}{128}q^4 - \frac{29}{2304}q^6 + \cdots$$

$$a_1\colon a = 1 + q - \frac{1}{8}q^2 - \frac{1}{64}q^3 - \cdots$$

$$b_1\colon a = 1 - q - \frac{1}{8}q^2 + \frac{1}{64}q^3 - \cdots$$

$$a_2\colon a = 4 + \frac{5}{12}q^2 - \frac{763}{13824}q^4 + \cdots$$

$$b_2\colon a = 4 - \frac{1}{12}q^2 + \frac{5}{13824}q^4 - \cdots.$$

McLachlan [1] has given a table of pairs of a and q that satisfy the characteristic numbers. These can be of value in creating an approximate stability diagram for the Mathieu equation.

Solutions of the type $ce_m(t, q)$ and $se_m(t, q)$ are referred to as Mathieu functions of order m and, in addition if m is integral as in the above, they are called functions of integral order.

6.1.2 Mathieu Functions of Fractional Order

If $a \neq m^2$, m integral, for $q = 0$, then the solutions obtained for q non zero will be of the form

$$ce_\nu(t, q) = \cos \nu z + \sum_{r=1}^{\infty} q^r C_r(t)$$

$$se_\nu(t, q) = \sin \nu z + \sum_{r=1}^{\infty} q^r S_r(t)$$

with

$$a = \nu^2 + \sum_{r=1}^{\infty} \alpha_r q^r.$$

It can be demonstrated that the characteristic number is the same for both $ce_\nu(t, q)$ and $se_\nu(t, q)$, as shown above, so that the solutions of both type can coexist and thus form a linearly independent pair. A complete solution therefore is of the form

$$x(t) = Ace_\nu(t, q) + Bse_\nu(t, q) \tag{6.8}$$

with A and B constants which will be determined by initial conditions.

Generally the functions $ce_\nu(t, q)$ and $se_\nu(t, q)$ are non-periodic, although bounded. They are therefore the solution types applicable in the stable regions of the Mathieu equation stability diagram. Even though the solutions are usually non-periodic it is instructive to put $\nu = m + \beta$, with m integral and $0 < \beta < 1$. If β can be expressed as p/s where p and s are integers with no common factors, then $ce_\nu(t, q)$ and $se_\nu(t, q)$ can be seen to be periodic with period πs or $2\pi s$ according as to whether p is even or odd.

6.1.3 Fractional Order Unstable Solutions

For values of a and q lying in an unstable region of the stability diagram the Mathieu functions of fractional order are of the form

$$ceu_\nu(\pm t, q) = e^{\pm \mu t}\phi(\pm t) \tag{6.9}$$

with $\phi(\pm t)$ periodic with period π or 2π. This type can be deduced from Eq. (2.25).

6.1.4 Limitations of the Classical Method of Treatment

Owing to the series nature of the solutions given by the classical approach, extensive calculation is generally required to ensure sufficient convergence to give accurate answers. Chapter 3 of McLachlan [1] however, does provide a range of techniques to assist in numerical evaluation of the classical methods, both for solutions and for characteristic numbers.

Determination of stability, classically, rests upon the use of the Hill determinant procedure developed in Sect. 4.3.6. Whilst the Hill determinant is strongly convergent for small a and q and thus, under those conditions, can give good estimates of characteristic exponents and stability, for values of a and q significantly larger than unity quite large determinants need to be calculated. Further, these may diverge initially prior to acceptable convergence. It is instructive to illustrate this.

Equation (4.25) shows that the characteristic exponent for the Mathieu equation can be obtained from

$$\mu = \frac{1}{\pi}\cosh^{-1}\left\{1 - 2\Delta(0)\sin^2\frac{\pi}{2}\sqrt{a}\right\} \tag{6.10}$$

since, for a sinusoidal coefficient $G_0 = a$. The determinant $\Delta(0)$ in the special case of the Mathieu equation is the tridiagonal form

$$\Delta(0) = \begin{vmatrix} \ddots 1 & \frac{-q}{a-16} & 0 & 0 & 0 \\ \frac{-q}{a-4} & 1 & \frac{-q}{a-4} & 0 & 0 \\ 0 & \frac{-q}{a} & 1 & \frac{-q}{a} & 0 \\ 0 & 0 & \frac{-q}{a-4} & 1 & \frac{-q}{a-4} \\ 0 & 0 & 0 & \frac{-q}{a-16} & 1 \ddots \end{vmatrix} . \quad (6.11)$$

Determination of a value for μ thus amounts to evaluation of $\Delta(0)$.

Inspection of Eq. (6.11) reveals an obvious difficulty. The determinant contains singularities at values of $a = (2r)^2$, r integral, and consequently cannot be evaluated at those values of a. Further these values influence convergence of the determinant, in the following manner. When evaluating $\Delta(0)$ a finite truncation such as 3×3, 5×5 etc. is used, relying upon the convergence of the determinant to ensure this is a reasonable approximation to the complete doubly infinite form. Such convergence will not occur however until the off-diagonal elements in Eq. (6.11) decrease rapidly. Clearly for a large, say around 10 this will not happen until sufficient rows and columns are included such that the denominators in the outer elements are approximately $(2r)^2$ ($(2r)^2 \gg a$) see Eq. (4.18) with $\mu = 0$, $\omega_p = 2$). For $a = 10$ in particular, up to an 11×11 determinant may be required in which the denominators of the off-diagonal elements in the top and bottom rows are $a - (2 \times 5)^2 = a - 100 \simeq 100$ (10% error). Table 6.1 shows values of $\Delta(0)$ when approximated by 3×3, 5×5, 7×7 and 9×9 finite truncations at several values of a and q, thereby illustrating the difficulty just highlighted. For a small (1.0) it is seen that convergence of the determinant is rapid and that a 3×3 truncation is sufficient to ensure reasonable accuracy. However for increasing a the effect of the singularities on convergence becomes evident. For $a = 3.8$, for example, reliable convergence does not occur until after the rows and columns with indices ± 2 are included (i.e. on the borders of a 5×5 determinant) and thus at least a 5×5 truncation is necessary. At $a = 10$ convergence is not apparent until at least a 7×7 determinant is used. In this case specific divergence is experienced as can be seen in the value of the 5×5 determinant. Divergence is also evident at $a = 30$, and in fact in this case convergence is only really noticeable in the 7×7 and 9×9 results.

Convergence of $\Delta(0)$ is also influenced by the magnitude of q since this parameter determines the relative strengths of the diagonal and off-diagonal elements (recall that the diagonal elements of $\Delta(0)$ are all unity). Table 6.2 shows this effect. Because of the increased significance of the remote elements with $q = 5.0$, convergence is considerably slower than with $q = 0.2$.

Clearly evaluation of stability by the classical method of Hill determinant analysis can be inaccurate unless sufficient precautions are taken to ensure

Table 6.1. Values of finite Hill determinants—illustrating the effect of numerical signularities

Size of determinant	a / q	1.0 / 0.2	3.8 / 1.0	10 / 1.0	30 / 1.0
3 × 3		1.02666	3.63160	0.96667	0.99744
5 × 5		1.02486	1.90131	1.02207	0.99196
7 × 7		1.02471	1.88768	1.00937	1.01578
9 × 9		1.02467	1.88573	1.00792	1.00596

Table 6.2. Values of finite Hill determinants—illustrating the effects of q

Size of determinant	a / q	1.0 / 0.2	1.0 / 5.0
3 × 3		1.02666	17.66670
5 × 5		1.02486	7.60474
7 × 7		1.02471	6.45606
9 × 9		1.02467	6.30290

reasonable convergence of the finite truncation chosen. It is also a time-consuming task owing to the need to evaluate very high order determinants. At best therefore the technique would only be considered for small values of a and small to moderate values of q. With a and q, for example, limited below 2, quite accurate $\Delta(0)$'s and thus values of the characteristic exponent can be obtained by hand calculation, with the assistance of a determinant evaluation routine in a programmable portable calculator. Fortunately for many of the applications of Mathieu equations such a range of a and q is appropriate and consideration may therefore be given to this technique.

Before proceeding it should be pointed out that the singularities which create the difficulties illustrated in Table 6.1 are only numerical. Mathematically, these singularities are removed by the zeros of the $\sin^2\frac{\pi}{2}\sqrt{a}$ term in Eq. (6.10) when determining a value for the characteristic exponent, and thus are not of significance theoretically.

Whilst the Hill determinant approach, within the cautions noted above, permits an approximation to stability to be determined, it does not, without significant additional computation, allow solutions to the Mathieu equation to be produced. Conversely, whilst the series expansion expressions of the preceding sections give approximate solutions, stability is not readily assessed by that technique. As a result it is clear that recourse is necessary to quite different philosophies for analysis. These include solution by numerical mathematical techniques in addition to the modelling approaches suggested in Chap. 5. These are pursued in the following sections.

6.2 Numerical Solution of the Mathieu Equation

Methods for numerical integration such as the well-known Runge-Kutta and Predictor Corrector techniques [2] lend themselves to equations with time-varying coefficients and thus can be used with the Mathieu equation. Within the limits of numerical accuracy, these procedures have the appeal that they give exact results since they do not rely upon algebraic approximations to the solutions of the equation. Their short-comings however lie with their numerical error, especially when solutions near stability boundaries are required. Accuracy is governed, by and large, by the number of iteration points chosen for the technique, per period of the periodic coefficient in the Mathieu equation. For solutions which are unreservedly stable or unstable, only a modest number of points—say 10 or 20 per period—is necessary to ensure reliable solutions. However near stability boundaries upwards of 60 points per period is necessary to give a solution accuracy of 0.1% after a total time interval corresponding to 50 complete periods of the time-varying coefficient.

Numerical solution of the Mathieu equation has been extensively used in the past [3, 4] to model systems which contain a sinusoidally time-varying parameter. A Fortran subroutine, based upon the Runge-Kutta Method in particular, has been given by Lever [5].

Whilst numerical methods are convenient tools with which to solve the Mathieu equation they have two disadvantages. First they are unable to give quantitative information regarding stability and secondly they are very time-consuming in evaluation.

6.3 Modelling Techniques for Analysis

The techniques presented in Chap. 5 are well-suited to the analysis of the Mathieu equation, owing to the very simple frequency spectrum of the sinusoidal coefficient. Following these procedures, a suitable combination of steps and ramps is sought as a replacement for the sinusoid and this combination is manipulated such that its lower frequency spectrum closely resembles that of the sinusoid. Simple Fourier analysis shows that the complex amplitude spectrum of a sinusoid is $|\Psi_{\pm 1}| = 1/2$, $\Psi_i = 0$ for all $i \neq \pm 1$. Consequently model coefficients should be chosen such that they have a spectrum which is, as nearly as possible, a single component of amplitude $\frac{1}{2}$ at the frequency of the sinusoid—i.e. at 2 radians per second (or a period of π). In addition of course, the viability of the method rests upon being able to solve analytically the resulting substitute equation.

Three likely models come to mind. First is that in which the sinusoid is replaced by a simple square waveform, the second is that in which a trapezoidal waveform is employed and the third model is that which utilises a staircase waveform. These are the subjects of the following sections.

6.3.1 Rectangular Waveform Models

The coefficients of the complex Fourier series of the rectangular waveform shown in Fig. 3.1 are given by

$$\Psi_m = \frac{1}{m\pi \mathrm{j}}[1 - \mathrm{e}^{-\mathrm{j}2m\tau}] \qquad m = 0, \pm 1, \pm 2, \ldots. \tag{6.12}$$

The general rectangular waveform has all frequency components. It is required

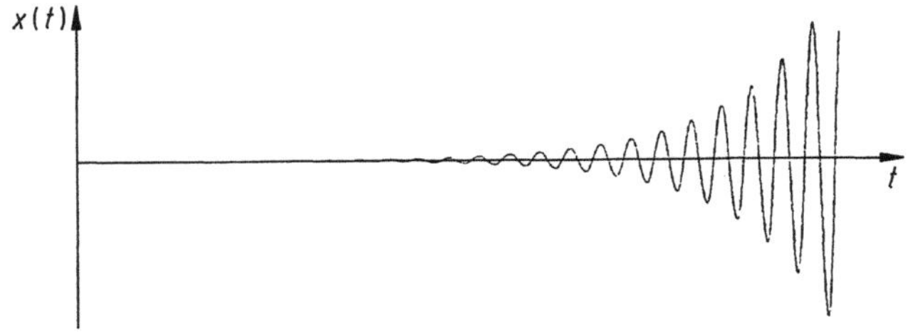

Fig. 6.1. Comparison of numerical and modelled solutions to the Mathieu equation at $a = 1.0$, $q = 0.1$, over 50 periods of the sinusoidal coefficient. Since agreement between the solutions is very good no distinction can be made in the diagram. The difference at the end of the time interval is about 0.15%; [6]

therefore to minimise as many of these as possible to cause its spectrum to resemble that of a sinewave. By choosing $\tau = \frac{\pi}{2}$—i.e. by making the waveform square—Eq. (6.12) shows that all even harmonics disappear, as is well known, leaving as a spectrum for the square waveform

$$|\Psi_{\pm 1}| = \frac{2}{\pi}, |\Psi_{\pm 3}| = \frac{2}{3\pi}, |\Psi_{\pm 5}| = \frac{2}{5\pi}, \ldots |\Psi_{\pm m}| = \frac{2}{m\pi}, m_{\text{odd}}.$$

This spectrum is fixed in its relative composition and thus the waveform must be used as it stands as a possible substitute for the sinusoid in the Mathieu equation, apart from an overall modification in amplitude. Since the period of the square waveform is π then its fundamental component needs to be matched to that of the sinusoid to provide the best possible model. As $|\Psi_{\pm 1}| = 1/2$ for the sinusoid the amplitude of the square wave must therefore be multiplied by $\pi/4$.

Owing to the significant third, fifth, seventh and ninth harmonic component content of the square wave it is to be expected that a poor model for the Mathieu equation would result. Whilst this is true near stability boundaries—i.e. for values of a and q that give rise to almost marginally stable solutions—it is not necessarily the case for values of a and q well away from boundaries. Applications therefore which do not depend upon conditions of marginal stability, such as degenerate parametric amplification, can be modelled very well. Figure 6.1 shows for example both modelled and actual solutions to a Mathieu equation with $a = 1.0$, $q = 0.1$. The Mathieu equation as such was solved using Runge-Kutta integration whereas the model equation employed a square waveform. As illustrated, the agreement between the two solutions is excellent demonstrating the adequacy of the model. The simulated solution was obtained in about 3% of the time required for the numerical procedure thereby highlighting a major attribute of the modelling approach.

6.3.2 Trapezoidal Waveform Models

For the general trapezoidal waveform shown in Fig. 5.3 but with $\theta = \pi$ to make it compatible in period with the Mathieu equation, the coefficients of the complex Fourier series are

$$\Psi_m = -\{1 - e^{-j2m\tau}\}\{1 - e^{-j2m\varepsilon}\}/2m^2\pi\varepsilon \qquad m = 0, \pm 1, \ldots \qquad (6.13)$$

Again this waveform can have all frequency components depending upon values given to τ and ε. If $\tau = \pi/2$ all even harmonics vanish, as is well known, leaving a complete set of odd components. However unlike the situation with a square waveform, the trapezoid has a further parameter (ε) which can be manipulated to minimise the higher frequency components, thereby permitting the spectrum of the trapezoid to resemble that of the sinusoid closely, simply by matching the fundamentals. In particular if ε is chosen as $\pi/3$ all harmonics which are multiples of 3 disappear including the third, and ninth, leaving the spectrum of the waveform as the fundamental, 4% of fifth harmonic, 2% of seventh harmonic and decreasing proportions of eleventh, thirteenth, etc harmonics. Provided the amplitude of the trapezoid is adjusted to 0.949703 to give a fundamental amplitude of 1/2, it is expected that the corresponding Hill equation would be a satisfactory substitute for the Mathieu equation. This is indeed the case, as demonstrated for a practical application in Sect. 7.1.5.

6.3.3 Staircase Waveform Models

Use of a staircase waveform approximation to the sinusoid in the Mathieu equation has the distinct advantage that the number of steps in the model waveform can be increased without limit in order to improve its resemblance to a sinewave. In view of the development in Chap. 5 it is, of course, not necessary to choose an extremely close geometric likeness to a sinusoidal waveform as such but simply one which models its lower frequency spectrum very well.

For a staircase model of the type depicted in Fig. 6.2, the step heights necessary to ensure the best possible (frequency domain) resemblance to a sinusoid are those given by Eq. (5.7). The spectral components of the waveform defined by these steps are related by the recurrence formula of Eq. (5.5). From these equations it can be deduced that, for a waveform with N equal time intervals, all harmonics of the model waveform are zero except those of the $(mN \pm 1)$th orders, where m is an integer; moreover these non-zero harmonics will have magnitudes of $1/(mN \pm 1)$ with respect to the magnitude of the fundamental. In particular the first non-zero harmonic above the fundamental is the $(N - 1)$th of relative amplitude $1/(N - 1)$. Thus N need only be chosen slightly larger than the highest harmonic considered to be significant in detracting from the model waveform's resemblance to a sinusoid.

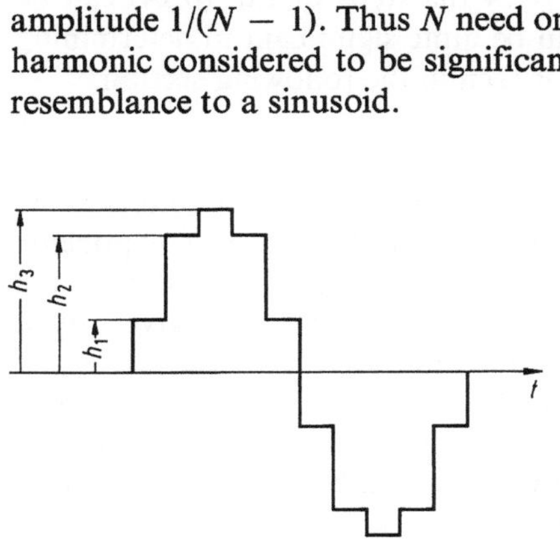

Fig. 6.2. Staircase model of a sinusoid with equal time steps.

Table 6.3. Step heights for staircase models of a sinusoid

Step heights (see Fig. 6.2)	number of steps 10	14	18	22	26
h_1	0.314159	0.224399	0.174533	0.142800	0.120831
h_2	0.822479	0.628753	0.502548	0.416831	0.355470
h_3	1.016640	0.908575	0.769948	0.657092	-0.569450
h_4	—	1.008442	0.944481	0.844120	0.750336
h_5	—	—	1.005096	0.962762	0.887615
h_6	—	—	—	1.003407	0.973309
h_7	—	—	—	—	1.002438
Next non-zero harmonic and its magnitude relative to the fundamental	$F_9 = \frac{1}{9}$	$F_{13} = \frac{1}{13}$	$F_{17} = \frac{1}{17}$	$F_{21} = \frac{1}{21}$	$F_{25} = \frac{1}{25}$

Table 6.3 shows the step heights for a range of staircase models for a sinusoid and indicates the degree of spectral impurity in each of those representations. Models of these types again find application in practice and, as with the simple square waveform, are rapid to evaluate by comparison to numerical integration of the Mathieu equation.

An alternative method for generating the step heights in a staircase model of the sinusoid in the Mathieu equation is to employ Eq. (5.8) which simply suggests that the step heights are the average of the sinusoid over the corresponding time interval. If N equal time intervals are chosen Eq. (5.8) leads to step heights of

$$h_n = \sin(2n-1)\pi/N \frac{\sin\frac{\pi}{N}}{\frac{\pi}{N}} \sin \qquad n = 1, \ldots, N. \tag{6.14}$$

Comparison of this expression with Eq. (5.7) shows that the heights given by this method differ from those based upon the more precise development of Sect. 5.2 by a factor of $[\sin(\pi/N)/(\pi/N)]^2$. Equation (6.14) will therefore only give step heights approximately the same as those of Eq. (5.7) when N is sufficiently large to cause $\sin(\pi/N)/(\pi/N) \to 1$. For $N = 14$ the steps of Eq. (6.14) can be "in error" by about 2.5%, an error which can be quite significant in determining values for the characteristic exponents. This is seen in the following section.

6.3.4 Performance Comparison of the Models

Model inadequacies show up most clearly when values for the characteristic exponent are calculated. For this reason computation of characteristic exponents has been used as a means for comparing the various models discussed in the preceding sections. Table 6.4 shows the characteristic exponents given by a number of models at several significant pairs of a and q. The "true values" of the exponents shown in the table were obtained using the Hill determinant procedure discussed in Sect. 6.1.4, employing 29×29 truncated determinants. Values for the staircase models shown in brackets are those given by deriving the models from Eq. (6.14) whereas the unbracketed entries were derived using Eq. (5.7). It

Table 6.4. Performance comparison of various models for the Mathieu equation [6]

a	3.67000^b	1.00000	-0.23330^b	0.23440
q	2.00000	0.10000	−0.70000	0.70365
True values[a]	0.0 + j0.01777	0.04994 + j0.0	0.01281 + j0.0	0.0 + j0.94782
Trapezoidal waveform model	0.0 + j0.01590	0.04994 + j0.0	0.01189 + j0.0	0.0 + j0.94789
10-step staircase model	0.0 + j0.01469	0.04994 + j0.0	0.00970 + j0.0	0.0 + j0.94806
	(0.0 + j0.05454)	(0.04832 + j0.0)	(0.13290 + j0.0)	(0.0 + j0.87278
14-step staircase model	0.0 + j0.01706	0.04994 + j0.0	0.01211 + j0.0	0.0 + j0.94786
	(0.0 + j0.04186)	(0.04911 + j0.0)	(0.09643 + j0.0)	(0.0 + j0.87995)
18-step staircase model	0.0 + j0.01752	0.04994 + j0.0	0.01255 + j0.0	0.0 + j0.94782
	(0.0 + j0.03472)	(0.04944 + j0.0)	(0.07578 + j0.0)	(0.0 + j0.91671)

[a] Computed using a 29 × 29 truncated Hill determinant.
[b] These points are very close to stability boundaries.

can be seen clearly by comparing these two sets of values to the "true values", that Eq. (6.14) leads to inaccurate models for small numbers of time intervals, especially in regions of marginal stability. The trapezoidal model is seen to perform acceptably even near stability boundaries. However as revealed in Chap. 3 these models demand substantial computational effort for their implementation.

6.4 Stability Diagrams for the Mathieu Equation

6.4.1 The Lossless Mathieu Equation

Stability diagrams are plots of a against q that depict regions of a and q for which the solution to a Mathieu equation is stable as opposed to those regions for which the solution is unstable. The regions of stability and instability are separated by *stability boundaries* on which the solution is marginally stable. If $\phi(\pi, 0)$ is the discrete transition matrix for the Mathieu equation computed, in principle, according to the material of section 2.3, then on stability boundaries trace $\{\phi(\pi, 0)\} = \pm 2$. The eigenvalues of $\phi(\pi, 0)$ are either both $+1$ or both -1 and, from Eq. (2.22), the characteristic exponent of the solution of the Mathieu equation is jm where m is an integer. As a result the fundamental solution to a Mathieu equation on stability boundaries is purely periodic. (As discussed in Sect. 4.5.7 there is a second, linearly independent solution that is a t-multiplied version of the first, owing to the degenerate eigenvalues.) In Sect. 6.1.1 it has been shown that purely periodic solutions occur for values of a and q related via equations such as Eq. (6.5) and Eq. (6.7)—in other words, for the equation's characteristic numbers. There are two sequences of characteristic numbers. One set—designated $a_0, a_1, a_2, \ldots$—correspond to even, periodic solutions of the equation whereas the other set—called $b_1, b_2, b_3, \ldots$—correspond to odd, periodic solutions. The first few are given in the penultimate paragraph of Sect. 6.1.1.

In view of the foregoing, to produce a stability diagram it is necessary only to

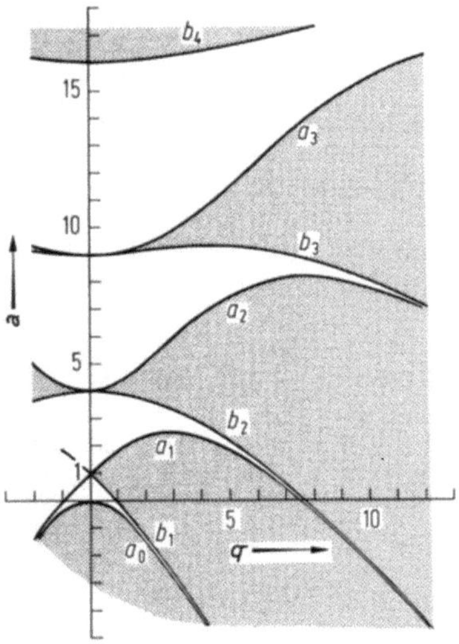

Fig. 6.3. Stability diagram for the Mathieu equation. Blank regions correspond to stable solutions and shaded regions to unstable solutions. The diagram, which is symmetrical about the a axis, was plotted from Appendix II of McLachlan [1]

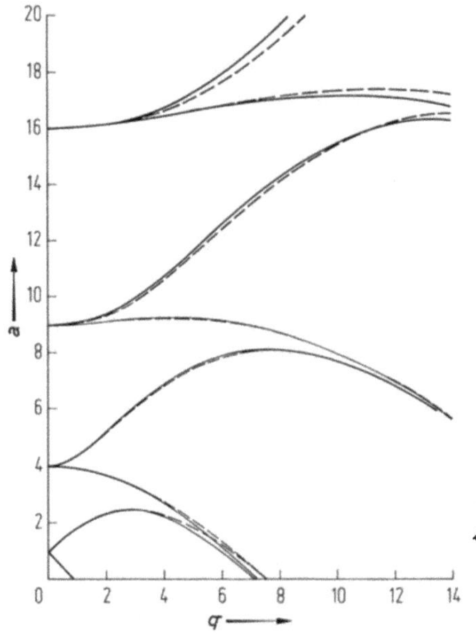

Fig. 6.4. Approximated stability diagram for the Mathieu equation. The real diagram has the boundaries shown in broken lines. (SIAM J. Appl. Math. (1976))

identify the a and q functional forms of the characteristic numbers in the region of (a, q) space of interest and then plot them. Since regions of instability are known to emanate from points on the a axis that are squares of integers it is easy to label the regions between the stability boundaries (see Sect. 4.3.7). The characteristic numbers have been computed for the standard Mathieu equation without a first derivative (loss) term, and McLachlan's Appendix II [1] tabulates a_0 to a_5 and b_1 to b_6. From these the stability diagram shown in Fig. 6.3 has been constructed.

Determination of the sets of characteristic numbers for the Mathieu equation and plotting them is the classical means by which the stability diagram has been determined. An alternative, and perhaps more convenient method, is to make use of the modelling technique outlined in Chap. 5. For this the sinusoid in the Mathieu equation is replaced by a sufficiently accurate counterpart that leads to an analytically tractable equation. The Meissner equation of Sect. 3.2 is such a candidate. A diagram for the substitute equation is then generated by computing and testing any of the characteristic exponent, the eigenvalues of the discrete transition matrix or simply the trace of the discrete transition matrix. An approximate or simulated, stability diagram for the Mathieu equation produced by this method is shown in Fig. 6.4. This is in fact the diagram for a Hill equation with a simple trapezoidal waveform coefficient such as that illustrated in Fig. 5.3 with ε chosen

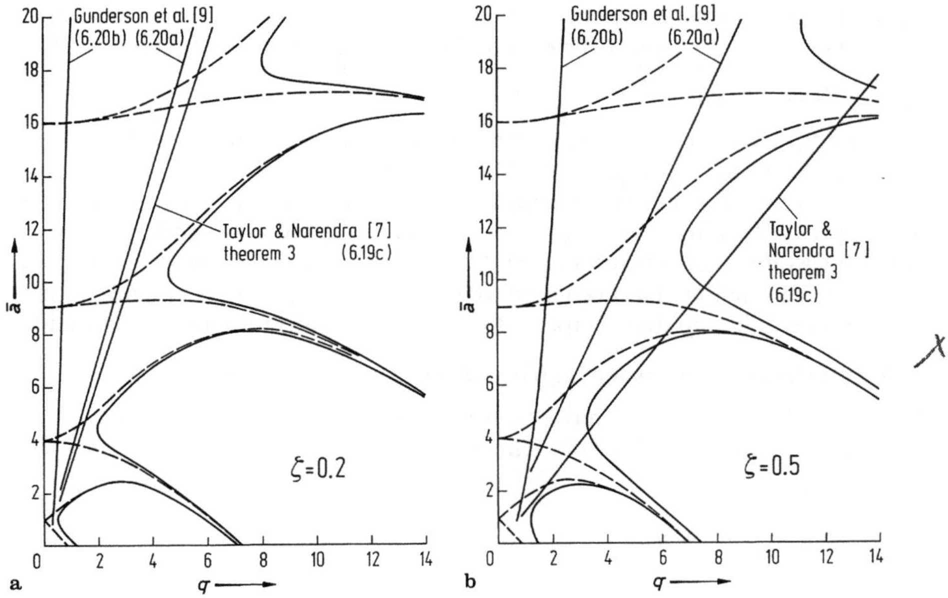

Fig. 6.5. Approximated stability diagrams for the lossy Mathieu equation with (a) $\zeta = 0.2$ and (b) $\zeta = 0.5$. Also shown are the stability criteria of Taylor and Narendra [7] and Gunderson et al. [9]. The broken curves represent the diagram for the lossless Mathieu equation. (SIAM J. Appl. Math. (1976))

as $\pi/3$ and τ taken as $\pi/2$ where π is the period of the waveform. With these values the frequency spectrum of the trapezoid resembles closely that of the sinusoid consisting of a fundamental, no harmonics that are multiples of two or three, and very small and diminishing fifth, seventh, eleventh etc. harmonics. The figure shows clearly that the simulated diagram is very accurate generally in the vicinity of the a axis and certainly for $a \lesssim q/2$. A more accurate simulated diagram is possible if the amplitude of the trapezoid is adjusted to 0.949703 to make its fundamental magnitude 0.5, thereby matching better the spectrum of the sinusoid in the Mathieu equation [6].

6.4.2 The Damped (Lossy) Mathieu Equation

Any real periodic system experiences some damping or loss. This shows up as a first derivative term to give a Mathieu equation of the form

$$\ddot{x} + 2\zeta\dot{x} - (\bar{a} - 2q\cos 2t)x = 0 \tag{6.15}$$

in which ζ is the damping constant. The transformation of Eq. (2.4), for ζ constant, is

$$y(t) = e^{\zeta t}x(t) \tag{6.16}$$

which when applied to Eq. (6.15) yields

$$\ddot{y} + (a - 2q\cos 2t)y = 0, \quad a = \bar{a} - \zeta^2 \tag{6.17}$$

If μ is the characteristic exponent of the solution to Eq. (6.17) then Eq. (6.16) shows that a necessary and sufficient condition for stability of $x(t)$ in Eq. (6.15) is

$$\mathrm{Re}\{\mu\} \leq \zeta \tag{6.18}$$

where the equality refers to the marginally stable case. This can be used to produce a stability diagram for the damped Mathieu equation in the following manner. For given values of $\bar{a}$ and q the characteristic exponent for Eq. (6.17) is computed from Eq. (2.27b) after having replaced 'cos 2t' by a suitable model such as a staircase waveform or a trapezoid. Stability of $x(t)$ is then assessed by applying Eq. (6.18). This method has been used to produce the diagrams of Fig. 6.5; in that case a simple unit amplitude trapezoid was used as a substitute for the sinusoid.

6.4.3 Sufficient Conditions for Stability of the Damped Mathieu Equation

Taylor and Narendra [7, 8] have determined three sufficient conditions for the solution to Eq. (6.15) to be stable. These are

$$q \lesssim \zeta\sqrt{a} \tag{6.19a}$$

$$q \lesssim \zeta\sqrt{a(1 + a)} \tag{6.19b}$$

and

$$q \lesssim \frac{\pi}{2} a\zeta \tag{6.19c}$$

with the restriction that $\zeta \ll 1$ and that $a \gg \zeta$ with a also of the order of 1 or greater. In Fig. 6.5 the third of these is shown Eq. (6.19c) to illustrate its validity.

Gunderson et al [9] have also introduced a pair of sufficient conditions for the stability of Eq. (6.15). These are

$$2q \leq (a - \zeta^2)\tanh \pi\zeta \tag{6.20a}$$

$$q - \zeta^2 < \zeta\sqrt{a} \tag{6.20b}$$

and are also illustrated in Fig. 6.5. In these conditions the assumption $\zeta \ll 1$ is not required. If it is used, it is seen that Eq. (6.20b) reduces to Eq. (6.19a). Hence Eq. (6.20b) is a general form of Eq. (6.19a).

Inspection of Fig. 6.5 shows that Eq. (6.20a) seems to be the most appropriate of the sufficient conditions for stability.

References for Chapter 6

1. Mc. Lachlan, N. W.: Theory and application of Mathieu functions. Clarendon: Oxford U.P. 1947. Reprinted by Dover, New York 1964
2. Ralston, A.: A first course in numerical analysis. New York: McGraw-Hill 1965
3. Dawson, P. H.; Whetten, N. R.: Ion storage in three dimensional, rotationally symmetric, quadrupole fields. I. Theoretical treatment. J. Vac. Sci. Technol. 5 (1968) 1–6
4. Dawson, P. H.; Whetten, N. R.: The monopole mass spectrometer. Rev. Sci. Instrum. 39 (1968) 1417–1422

5. Lever, R. F.: Computation of ion trajectories in the monopole mass spectrometer by numerical integration of Mathieu's equation. IBM J. Res. Dev. 10 (1966) 26–39
6. Richards, J. A.: Modelling parametric processes—a tutorial review. Proc. IEEE 65 (1977) 1549–1557
7. Taylor, J. H.; Narendra, K. S.: Stability regions for the damped Mathieu equation. SIAM J. Appl. Math. 17 (1969) 343–352
8. Narendra, K. S.; Taylor, J. H.: Frequency domain criteria for absolute stability. New York: Academic Press 1973
9. Gunderson, H.; Rigas, H.; van Vleck, F. S.: A technique for determining stability regions for the damped Mathieu equation. SIAM J. Appl. Math. 26 (1974) 345–349

Problems

6.1 The a_0 stability boundary for the Mathieu equation is

$$a = -\frac{1}{2}q^2 + \frac{7}{128}q^4 - \frac{29}{2304}q^6 + \cdots$$

Show that for q small this is the same as the expression in Prob. 5.5 derived from a truncated Hill determinant.

6.2 The upper left hand entry in Table 6.3 is 0.1π. Verify this from Eq. (5.7).

6.3 Use a 3×3 truncated Hill determinant, along with Eq. (6.10) to determine values of the characteristic exponent at the a, q points given in Table 6.4.

6.4 Show that the equation

$$\ddot{x} + 2\zeta\dot{x} + ax = 0$$

is always stable for $a, \zeta > 0$, (consider Eq. (6.15)) and always unstable for $\zeta < 0$.

6.5 Hill determinant procedures can be used to determine stability of the lossy Mathieu equation in Eq. (6.14). Repeat the procedure of Sect. 4.3.6 to derive the Hill determinant and show that this leads to the same result as adopting Eq. (6.15) first and using the determinant for the lossless equation.

6.6 Consider the following Mathieu equation with a first derivative

$$\ddot{x} + \gamma\cos 2t\dot{x} + (a - 2q\cos 2t)x = 0$$

Show that it can be transformed to

$$\ddot{y} + \left\{\left(a - \frac{\gamma^2}{8}\right) - 2q\cos 2t + \gamma\sin 2t - \frac{\gamma^2}{8}\cos 4t\right\}y = 0.$$

Suppose $\gamma = 0.1$ so that it can be approximated as

$$\ddot{y} + \{a - 2q\cos 2t + \gamma\sin 2t\}y = 0.$$

Show how a suitable staircase waveform model can be devised for any range of q and γ.

Part II
Applications

Chapter 7

Practical Periodically Variable Systems

Systems described by periodic differential equations arise in practice in a variety of unrelated fields. These may be systems with parameters that vary periodically with time or with some other independent variable such as position, or else they may be field-like situations with boundary conditions that lead to periodic equations upon separating variables. In this chapter a representative cross section of applications is treated with particular emphasis on those of major importance. With regard to the Mathieu equation, in particular, McLachlan [1] has dealt with a range of practical illustrations such as electrical and thermal diffusion in systems with elliptical boundary conditions, elliptic waveguides, diffraction of acoustic and electromagnetic waves around ellipsoidal objects and mechanical vibrations. He has also provided treatments, in terms of Mathieu functions, of frequency modulation and the behaviour of loudspeakers. Some of these same applications are treated here. In addition however more recent illustrations of periodically parametric behaviour are covered, with emphasis on analysis, particularly from a computational viewpoint.

7.1 The Quadrupole Mass Spectrometer

Electric and magnetic lens structures such as those shown in Fig. 7.1.1 are commonly used for focussing beams of charged particles. In particular the magnetic quadrupole is used in synchrotron particle accelerators to contain beams against the defocussing effect of space charge [2]. In 1953 Paul and Steinwedel [3]

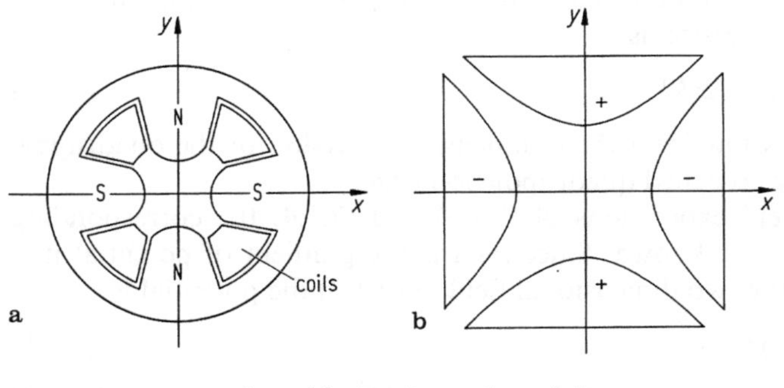

Fig. 7.1.1. a Magnetic and **b** electric quadrupole lenses

proposed that an electric quadrupole can be used for mass selection if it is energised by a high frequency electric potential and ions are injected axially through the structure. Selectivity is obtained from the dynamic (in fact parametric) interaction of the ions and the field and, as a consequence, such an arrangement is said to belong to the class of dynamic mass spectrometer [4]. This is to distinguish it from static instruments such as those that discriminate masses by particle injection into a constant magnetic field and those that make use of the differing field free drift velocities of monoenergetic ions of different mass [5]. The latter are referred to as time-of-flight instruments whereas the former are the commonly encountered magnetic mass spectrometers.

In this section the varieties of mass spectrometer based upon a quadrupole lens are outlined to illustrate how periodic equations arise in a description of their principles of operation. Only sufficient practical detail as is of relevance to the theme of this book is given. For a comprehensive account of all aspects of the quadrupole mass spectrometer and its derivatives—the three dimensional ion cage and the monopole mass spectrometer—the reader is referred to Dawson [6].

7.1.1 Spatially Linear Electric Fields

Ideally, the electric field employed by quadrupole mass spectrometers is linear with position. In three cartesian coordinates therefore

$$\boldsymbol{E} = \delta\{a_1 x\boldsymbol{i} + a_2 y\boldsymbol{j} + a_3 z\boldsymbol{k}\} \tag{7.1.1}$$

where $\boldsymbol{i}$, $\boldsymbol{j}$ and $\boldsymbol{k}$ are unit vectors in the x, y and z directions of the coordinate space; a_1, a_2 and a_3 are weighting constants and δ is a position independent factor that may be a function of time.

Laplace's equation requires $\nabla \cdot \boldsymbol{E} = 0$. When applied to (7.1.1) it yields

$$a_1 + a_2 + a_3 = 0. \tag{7.1.2}$$

To specify the field fully an arbitrary choice needs to be made in (7.1.2). For example, if a_3 is put to zero $a_2 = -a_1 = -a$ in which case the field becomes

$$\boldsymbol{E} = \delta a\{x\boldsymbol{i} - y\boldsymbol{j}\}. \tag{7.1.3}$$

This is the spatial form of the field conventionally adopted in the quadrupole mass spectrometer. If, on the other hand, $a_3 = -(a_1 + a_2) = -2a$, with $a_1 = a_2 = a$, then the field expression is

$$\boldsymbol{E} = \delta a\{x\boldsymbol{i} + y\boldsymbol{j} - 2z\boldsymbol{k}\} \tag{7.1.4}$$

which is the field expression for a three dimensional version of the quadrupole mass spectrometer—the so-called quadrupole ion cage.

To implement the field expressions of (7.1.3) and (7.1.4) the corresponding scalar potentials need to be known. Since field is the gradient of potential it is shown readily that for the two dimensional field of (7.1.3) the potential is

$$\phi = \tfrac{1}{2}\delta a\{x^2 - y^2\} + c \tag{7.1.5}$$

where c is a constant that is generally set to zero. The equipotentials of a two dimensional linear electric field therefore are sets of rectangular hyperbolas. To

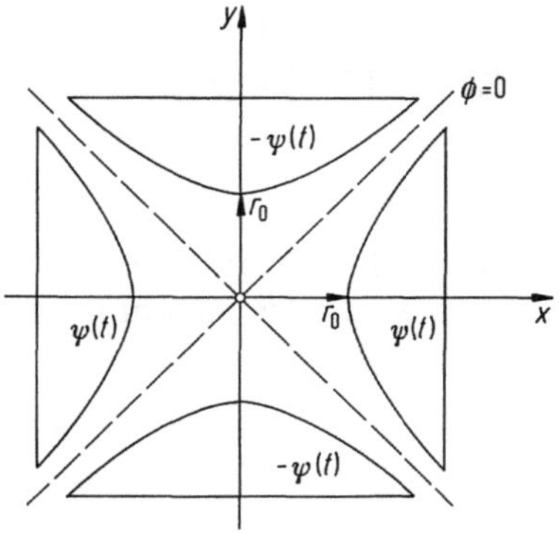

Fig. 7.1.2. Electrode structure and applied potentials in a quadrupole mass spectrometer

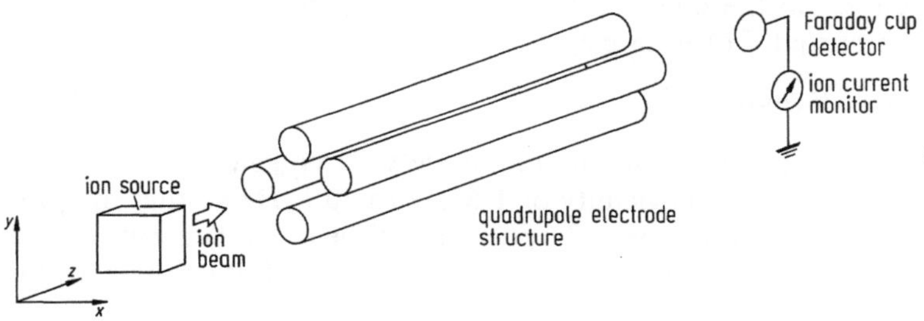

Fig. 7.1.3. Components of a quadrupole mass spectrometer

generate the field it is sufficient then to choose a set of electrodes that conform to particular equipotentials. This requires that hyperbolic cross section electrodes (ideally of infinite length but finite in practice) be used as shown in Fig. 7.1.2. If opposite electrodes are placed $2r_0$ apart, where r_0 is referred to as the 'field radius' and if $\pm\psi(t)$ is the actual electric potential applied to those electrodes it can be seen that $\delta a = 2\psi(t)/r_0^2$ so that

$$\phi(t) = \psi(t)\{x^2 - y^2\}/r_0^2 \tag{7.1.6a}$$

and

$$\boldsymbol{E}(t) = 2\psi(t)\{x\boldsymbol{i} - y\boldsymbol{j}\}/r_0^2. \tag{7.1.6b}$$

Clearly, as electrodes with hyperbolic cross section are difficult to make and align in practice, cylindrical rods with suitably chosen radii are employed. It has been shown that if rods with a radius of 1.148 r_0 are used, the field generated is linear to within $\pm 1\%$ over the central 80% of the region enclosed by the electrodes [7].

7.1.2 The Quadrupole Mass Filter

The components of a quadrupole mass spectrometer are shown in Fig. 7.1.3. Ions are injected into the quadrupole electrode structure with constant energy in the z direction and are detected at the output end by a simple Faraday cup or by a cup preceded by a secondary electron multiplier to provide current gain. Since there is no field in the z direction the ions drift axially through the quadrupole field with constant velocity as determined by their injection energy. In the transverse xy plane however their motion is described by

$$m\ddot{\boldsymbol{r}} + e\boldsymbol{E} = 0 \tag{7.1.7}$$

where $\boldsymbol{r}$ is the position vector $x\boldsymbol{i} + y\boldsymbol{j}$ and m and e are the ion's mass and charge respectively. Together (7.1.6b) and (7.1.7) give the separated equations of transverse ion motion

$$\ddot{x} + 2e\psi(t)x/mr_0^2 = 0 \tag{7.1.8a}$$

$$\ddot{y} - 2e\psi(t)y/mr_0^2 = 0. \tag{7.1.8b}$$

In the following it will be seen that mass discrimination results if the energising potential $\psi(t)$ is chosen as the sum of a dc voltage and a periodically time-varying voltage such as a sinusoid. For example

$$\psi(t) = U + V\cos\omega t \tag{7.1.9}$$

where U is the dc term and V and ω are the amplitude and frequency of the periodic term. Usually $f = \omega/2\pi$ is in the vicinity of 1 MHz to 10 MHz so that the time variation is often referred to as the "rf" potential. When (7.1.9) is substituted into (7.1.8) a pair of periodic differential equations result. Additionally with the change of time variable

$$2\xi = \omega t$$

both of these equations can be expressed in the canonical form of the Mathieu equation

$$\ddot{x} + (a - 2q\cos 2\xi)x = 0.$$

For motion in the x direction

$$a_x = \delta eU\eta \quad \text{and} \quad q_x = -4eV\eta \tag{7.1.10a}$$

with $\eta = (mr_0^2\omega^2)^{-1}$, whilst for motion in the y direction the Mathieu equation has

$$a_y = -\delta eU\eta \quad \text{and} \quad q_y = 4eV\eta. \tag{7.1.10b}$$

The advantage of expressing the pair of equations in the form of a simple Mathieu equation is that the well-known properties of the Mathieu equation can be exploited to describe the operation of the instrument. In particular the equation's stability diagram can be examined to see how the transverse ion motion is determined by the values of a_x, a_y, q_x and q_y defined in (7.1.10).

As a result of practical considerations, the only portion of the Mathieu equation stability diagram generally relevant to the operation of the quadrupole mass spectrometer is that in the vicinity of the first stable region about the origin. This is shown in Fig. 7.1.4. Examination of (7.1.10) shows that the behaviour of ions in transit through the quadrupole structure can be described by a pair of points in the stability diagram lying in the second and fourth quadrants. They

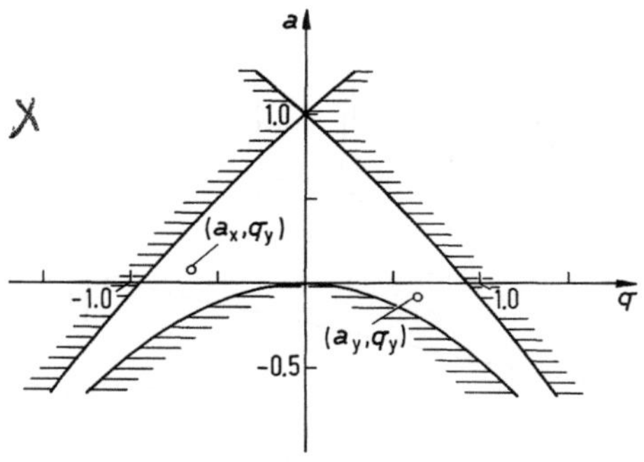

Fig. 7.1.4. Regions of Mathieu equation stability diagram relevant to mass filter operation

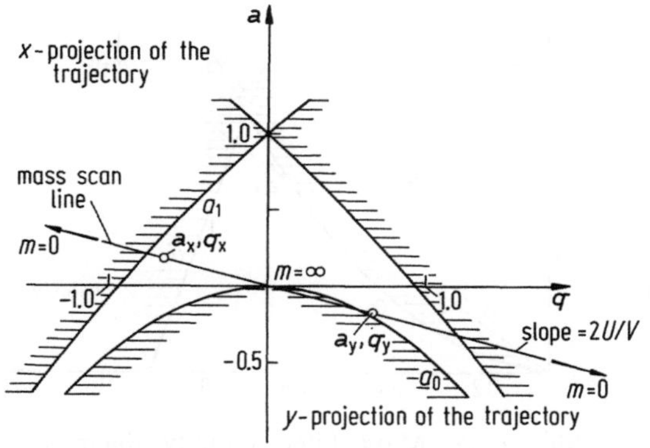

Fig. 7.1.5. Stability diagram with the mass scan line illustrated

are reflections of each other in the origin. If both points fall in a stable zone as depicted in Fig. 7.1.4 the ionic trajectory, corresponding to the energising potential defined by U, V and ω and for an ion of mass m, will be classically stable in both the x and y components so that the ion has a high probability of traversing the length of the quadrupole structure without collision with the electrodes. It will thus be detected and counted. If the operation of the device depended wholly upon classical stability, as implied in this description, the probability of ion transmission would be exactly unity. However the finite length of the quadrupole electrode structure (and thus finite time of flight of an ion) and the amplitude bound established by the electrode placement impose short time stability constraints upon both components of the ion trajectory. In many cases this reduces the probability of ion transmission to something less than unity.

If either of (a_x, q_x) and (a_y, q_y) as defined in (7.1.10) fall in an unstable zone the corresponding ion trajectory component grows without bound leading the ion to strike one of the electrodes whereupon it becomes neutralised and is no longer influenced by the field. Consequently it will not reach the ion detector and will not be counted.

The definitions of (a_x, q_x) and (a_y, q_y) in (7.1.10) show that for a given fixed excitation potential there will be as many pairs of (a, q) points as there are ions of different mass being injected into the quadrupole electrode structure from the ion source. In fact a line can be constructed, referred to as the mass scan line, that displays the actual distribution as those operating points with mass. This line, shown in Fig. 7.1.5, is symmetric about the origin with the heavier masses lying towards the origin and the lighter masses tending towards higher values of (a_x, q_x) and (a_y, q_y). It is evident therefore that the y component of the ion trajectory acts to reject higher mass number ions whereas the x component rejects lighter ions. Together they determine a band of ion mass passed by the structure. Because of the analogy of mass selectivity here to the frequency selectivity of an electric circuit filter, the quadrupole electrode arrangement, and frequently the complete instrument, is called a *mass filter*.

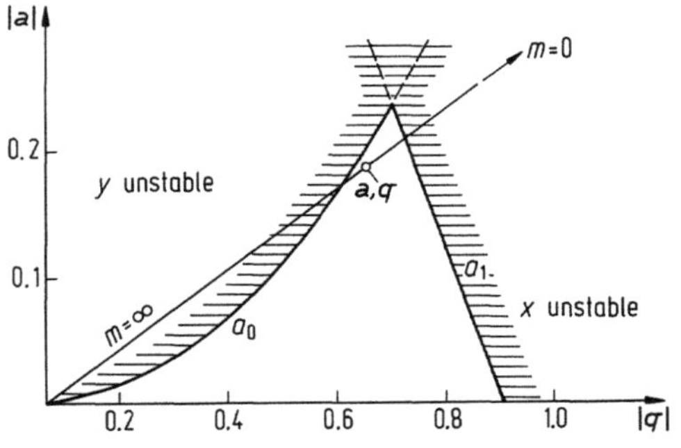

Fig. 7.1.6. Mass filter stability diagram $a = |a_x|, |a_y|, \quad q = |q_x|, |q_y|$.

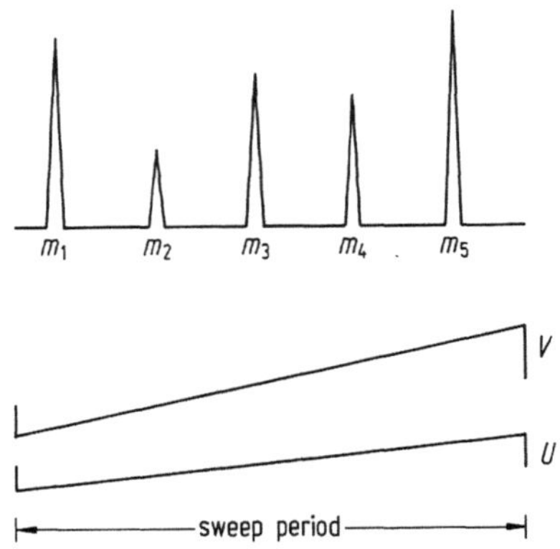

Fig. 7.1.7. Typical mass spectrum from a quadrupole mass filter showing the swept dc and rf voltages for comparison

Since the mass scan line passes through and is symmetrical about the origin of the stability diagram, considerable insight into the operation of the mass filter is given if the second and fourth quadrants are superimposed and replotted as shown in Fig. 7.1.6 to form the mass filter stability diagram. A point on the new (common) mass scan line now defines the properties of both projections of an ionic trajectory. The stable portion of the diagram—often referred to as the stability triangle—represents those operating points for which both the x and y components of a trajectory, and thus the trajectory as a whole, is stable. Outside the triangle either one or both of the components yields a trajectory unstable.

With a suitable choice of the slope of the mass scan line (which is $a/q = 2U/V$), such that it intersects the stability triangle just below its apex, an extremely selective mass filter can be obtained. In other words the filter can be tuned electrically by the relative choice of U and V so that only a narrow band of atomic mass number (which can be less than 1 amu) will be transmitted by the filter. To alter the central mass in this narrow band both the dc and rf voltage amplitude need to be changed whilst maintaining their ratio constant to ensure operation with the same mass band width (referred to as selectivity or resolution). These voltages are varied continuously in a sawtooth fashion to allow a mass spectrum to be scanned such as that shown in Fig. 7.1.7. The spread of each peak is a result of some ions successfully traversing the filter even though the voltages are such that the corresponding (a, q) point for those ions is not *exactly* in the centre of the stable portion of the stability triangle. This is a result, in part, of the dependence of the device upon *short time* stability.

Because of its simplicity, compactness and ruggedness the quadrupole mass spectrometer has remained an attractive analytical tool since it became commercially available in the early 1960's. Performance has been steadily improved and the device has been linked to other instruments such as gas chromotographs to provide highly versatile systems for compound analysis. Usually these incorporate computers for control and automated analysis of results.

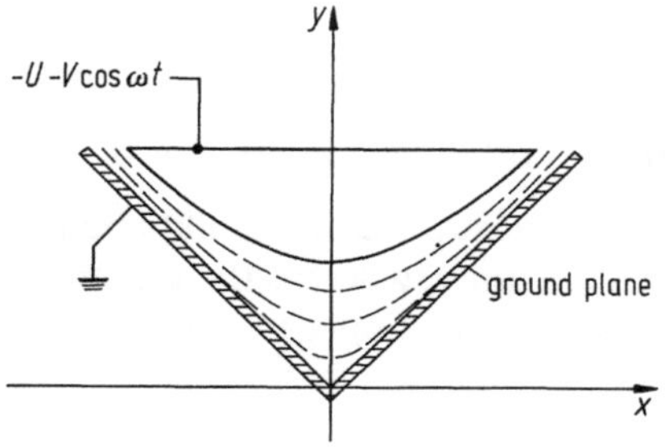

Fig. 7.1.8. Electrode configuration for the monopole mass spectrometer

Variations on the basic quadrupole device have been proposed. Some relate to its geometry and are mentioned briefly in the following two sections. Others however relate to the periodic energising potential. In particular it has been proposed that practical and analytical advantages arise if the combined dc and sinusoidal rf potential are replaced by a single rectangular rf waveform that incorporates its own dc component intrinsically. Such a mode of operation, which has been demonstrated both experimentally and by computer modelling [8] is based upon the Meissner equation. The arbitrariness of the time-varying portion of the excitation potential, leading to a general Hill equation description of the instrument, has also been demonstrated [9, 10], the essential requirement for selective mass filtering being that the potential possess a dc term, either externally added or intrinsic to the waveform. Notwithstanding this generality, only sinusoidally and rectangularly operated devices have been constructed.

7.1.3 The Monopole Mass Spectrometer

Equation (7.1.6a) reveals that there are zeros of potential in the quadrupole field along the lines $|x| = |y|$. Consequently an equivalent quadrupole field can be synthesised by the single electrode and earthed rightangled structure shown in Fig. 7.1.8. Images of the single electrode in the angle give rise to a virtual quadrupole. When used in place of a real quadrupole for mass selection the corresponding instrument is called a monopole mass spectrometer. It was first proposed and demonstrated experimentally by von Zahn [11] in 1963.

The electric field expression in the monopole mass spectrometer is given by

$$\boldsymbol{E}(t) = 2\psi(t)\{x\boldsymbol{i} - y\boldsymbol{j}\}/r_0^2, \quad y \geq 0, |x| \leq y \tag{7.1.11}$$

which, by reason of the constraints, indicates that the potential applied to the energised electrode is that which is applied in the y direction of a quadrupole field. This is necessary since the dc term has to be negative to create a trajectory that does not attempt to cross the quadrupole axis at a rate similar to the field frequency (as it does for an x component) but rather remains above the axis and thus in the limited monopole region for a large number of rf cycles. This difference in x and y components of an ion trajectory is illustrated in Fig. 7.1.9 and can be explained somewhat by noting from McLachlan [1] that the solution of a Mathieu (or Hill) equation in a stable region can be expressed generally as

$$x(t) \text{ or } y(t) = A \sum_{r=-\infty}^{\infty} c_{2r} \cos(2r + \beta)t + B \sum_{r=-\infty}^{\infty} c_{2r} \sin(2r + \beta)t$$

$$= Ac_0 \cos \beta t + Bc_0 \sin \beta t + \text{other terms with } r \neq 0.$$

The x solution in a quadrupole-like instrument is close to the a_1 stability boundary for which the characteristic exponent β is close to unity, whereas the y component is close to the a_0 boundary for which β is close to zero. Consequently

$$x(t) \simeq Bc_0 \sin t + \text{other terms}$$

whereas

$$y(t) \simeq Ac_0 \cos \beta t + \text{other terms}, \quad \beta \simeq 0.$$

The first term in $x(t)$ is a sinuoid which is subharmonic to the rf potential and causes the x solution to cross the time axis (quadrupole axis) at least every second period of the excitation. On the other hand the first term of $y(t)$ is a very slowly varying additive sinusoid which lifts the entire solution off the axis as depicted in Fig. 7.1.9. A completion of this line of reasoning requires a consideration of the amplitude of the remaining terms compared with Ac_0. Nevertheless the trend of the x and y solution types is evident.

The geometric field constraints in the monopole structure make the device dependent upon spatial focussing at an exit aperture in addition to being dependent

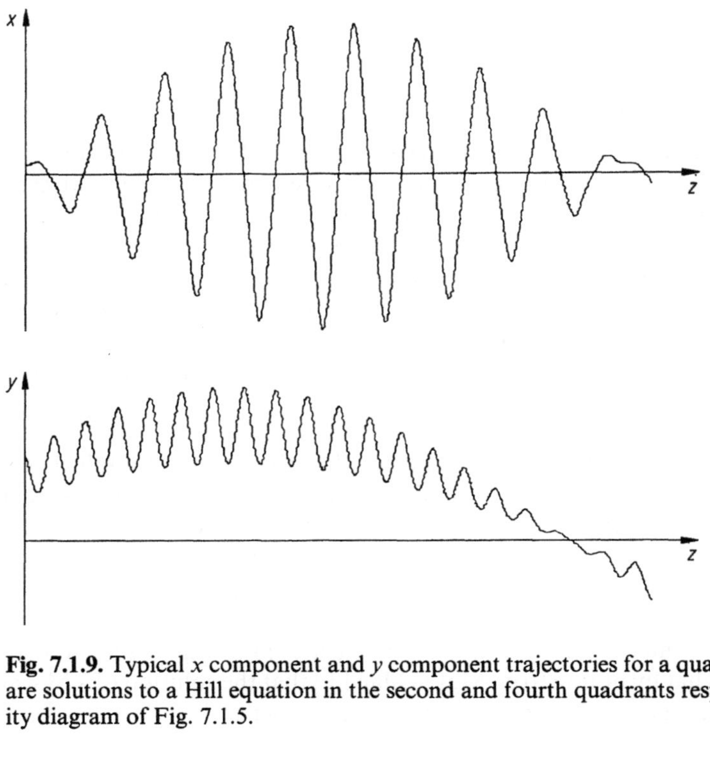

Fig. 7.1.9. Typical x component and y component trajectories for a quadrupole device. These are solutions to a Hill equation in the second and fourth quadrants respectively of the stability diagram of Fig. 7.1.5.

upon conditions of stability and instability of the ion trajectory. Consequently its mode of operation is slightly different to that of the quadrupole filter. A discussion of these effects and a fuller description of the operation and performance of the monopole mass spectrometer is given in Hudson and Watters [12] and Dawson [6].

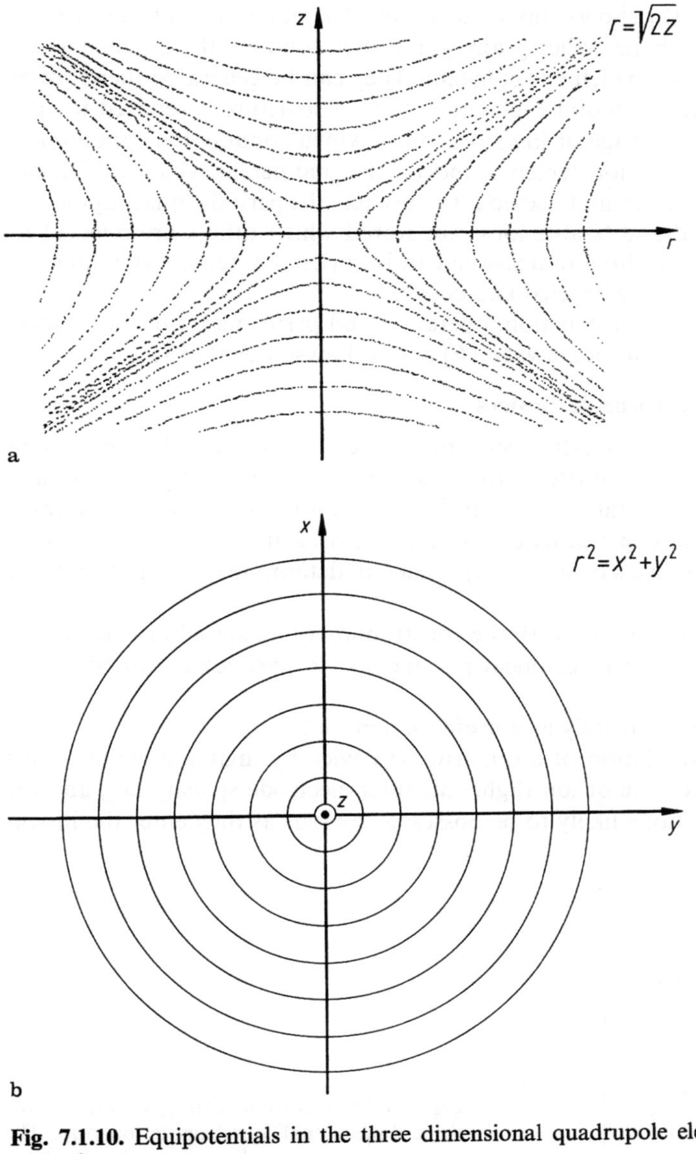

Fig. 7.1.10. Equipotentials in the three dimensional quadrupole electric field. **a** *rz* plane; **b** *xy* plane

7.1.4 The Quadrupole Ion Trap

Integration of (7.1.4) shows that the equipotentials of a three dimensional linear electric field are described by

$$\phi = \tfrac{1}{2}a\delta\{x^2 + y^2 - 2z^2\}$$

which, in the cylindrical coordinates (r, z) where $r^2 = x^2 + y^2$, is

$$\phi = \tfrac{1}{2}a\delta\{r^2 - 2z^2\}. \tag{7.1.12}$$

Examination of (7.1.12) shows this field to have the equipotentials depicted in Fig. 7.1.10. They are hyperbolic in any plane that contains the z axis and are circular in all planes for which z is constant. They can be generated therefore by the electrode structure shown in Fig. 7.1.11. Such a structure is referred to variously as an ion trap, ion cage or ion store, all of which connote ion containment rather than ion transmission which is indeed the manner in which this device operates. Ions of 'all' mass are injected into the cage, the periodic quadrupole-like potentials are applied and 'stable ions' are stored whilst others are rejected by virtue of their unstable three dimensional trajectories. At some latter time the stable, contained ions are extracted and counted.

Further details of the construction, operation and performance of the quadrupole ion cage will be found in Dawson [6] and in Lawson et al [13].

7.1.5 Simulation of Quadrupole Devices

A convenient means by which the operating characteristics of the three variations of quadrupole mass spectrometer may be assessed is to use a digital computer model. This entails computation of the trajectories followed by ions in the devices and the application of constraints on those trajectories imposed by the relevant geometries. The advantages of such an approach to determine device performance are that

(i) it is inexpensive compared with the construction of an actual instrument,
(ii) it is convenient since mechanical parameters, in particular, can be varied readily, and
(iii) the model serves ultimately as a useful design tool.

To be effective, simulation of a quadrupole device by digital computer must account for the finite time of ion flight, the inter-electrode spacing $2r_0$, and the range of initial conditions likely to be possessed by ions as they enter the instru-

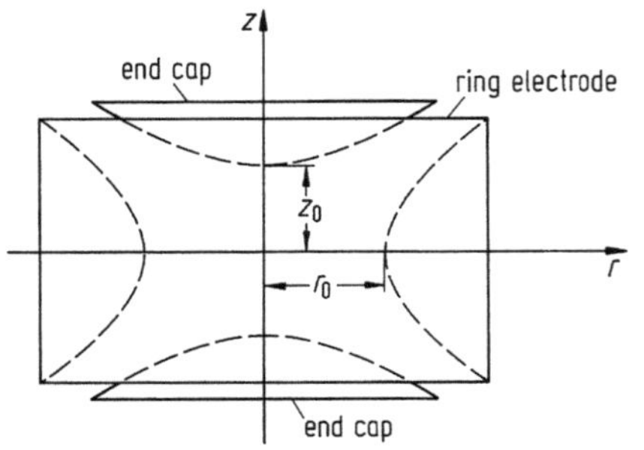

Fig. 7.1.11. Electrode configuration for the three dimensional, quadrupole ion trap. The structure is rotationally symmetric about the z axis

ment. These initial conditions include initial displacement and transverse velocity in both the x and y trajectory components, in addition to the phase of the rf field component at the instant of ion injection. Typically therefore simulation involves solution of the Mathieu equations which describe the two trajectory projections with the five initial conditions for a particular ion and checking that the trajectory amplitudes do not exceed the spacing between electrodes. Often the initial transverse velocities are assumed to be zero thus leaving only three ion initital conditions to be considered. Nevertheless a model which represents adequately the full range of injection phase, a reasonable distribution of initial displacements and a representative number of ions injected over the time required to produce one peak shape, on the face of it will demand around 100,000 effective solutions of a Mathieu equation. Consequently the solution technique has to be highly efficient or else extensive computational capacity must be available. By assuming zero initial transverse velocities in each component however (i.e. that the ion beam is well collimated) the number of solutions to the Mathieu equation actually required is reduced to around 500 to 1000 since solutions for only one initial x and y displacement is necessary. Trajectories for other displacements can then be inferred by simple scaling.

Digital computer models for a quadrupole device were first used by Lever [14] and Dawson and Whetten [15, 16] to determine the operating characteristics of a monopole mass spectrometer and the quadrupole ion cage [17]. These models were limited in that only a single initital ion displacement was employed and, in the case of a monopole, only the y trajectory component was simulated. This last restriction does not represent a severe departure form accuracy since the x trajectory component in a monopole under normal operating conditions is of small amplitude and has little influence on device selectivity. In these treatments the Mathieu equations describing ion motion were solved by numerical integration methods and consequently were time consuming.

Improved models for quadrupole devices, based upon the Mathieu equation, were developed following the proposal for a device operated with a rectangular rf waveform [8, 18]. One advantage of that mode of operation is that the equation describing ion motion—the Meissner equation—is analytically tractable, as demonstrated in Sect. 3.2. Consequently it was realised, along the lines suggested by Pipes [19], that the sinusoidal rf term in the conventional operating mode could be represented by a staircase approximation, such as that shown in Fig. 7.1.12, and therefore could be solved exactly. This would lead to rapid device

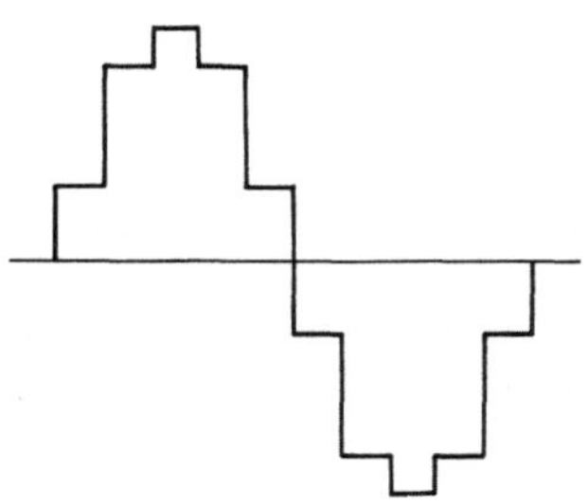

Fig. 7.1.12. Staircase model of a sinusoid, with equal time steps

simulation since numerical integration methods are not required. This was first used for the simulation of sinusoidally operated quadrupole devices by Dawson [20, 21] who employed a 200 segment approximation to a sinusoid over a single period. However in view of the demonstrated lack of significance of the higher harmonics of the rf waveform on the operation of a quadrupole instrument [22], it was shown shortly thereafter that as few as 14 or so steps could be sufficient to represent the sinusoid with, provided the step heights were chosen according to the modelling philosophy of Chap. 5 [23]. As a result very rapid computer models for ideal conventional quadrupole and monopole mass spectrometers, and ion cages, are now possible.

An alternative model of a similar style was proposed in 1975; in this, the sinusoid is represented by a trapezoid [24]. Again, this is a rapid model compared with simulation based upon numerical integration of the Mathieu equation (by about an order of magnitude) owing to the tractability of the model equation as seen in Sect. 3.9. It yields results however that resemble very closely those obtained by numerical integration. Fig. 7.1.13 for example shows simulated peak shapes from both approaches in two ranges of resolution (i.e. selectivity) demonstrating the accuracy of the fast trapezoidal model.

More recent quadrupole device modelling investigations have been performed using the phase-space method of solution outlined in Sect. 3.15. Typical of these are the studies reported by Bonner et at [24a], Baril [24b], Dawson [24c], Todd, Freer and Waldren [24d, 24e] and Chun-Sing and Schwessler [24f].

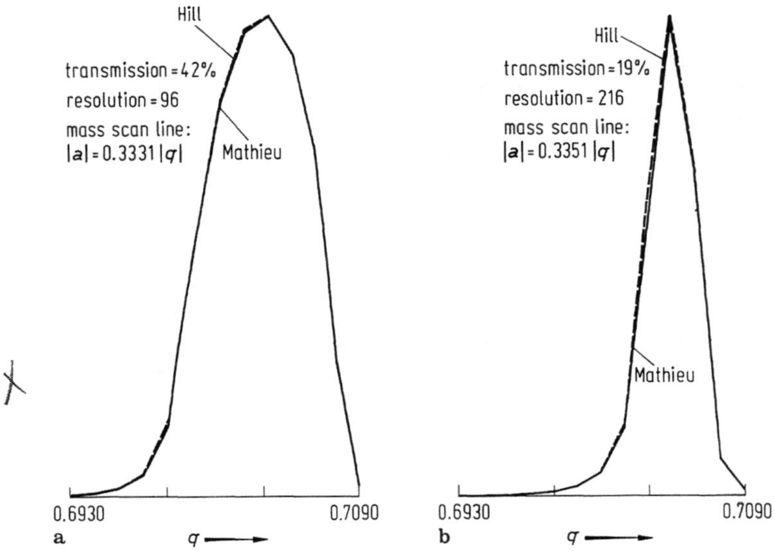

Fig. 7.1.13. Peak shapes for a quadrupole mass filter produced by computer modelling of the device. The full curves are those based upon numerical solution of the Mathieu equation whereas the broken curves are those given by using a trapezoidal model for the sinusoid in a Hill equation equivalent. Even at high selectivities (resolution) the modelled results agree well with those from numerical integration [24]

7.1.6 Non idealities in Quadrupole Devices

The description of the operation of quadrupole devices by Mathieu equations is accurate only when the instrument has ideally shaped hyperbolic electrodes. For a real instrument, in which machining irregularities exist in the manufacture of the electrodes or in which circular cylindrical rods are employed, the Mathieu equation description is only an approximation, albeit a sufficiently accurate one for many purposes. If detailed consideration of device operation is required it is necessary to resort to better descriptions. When the geometry of the device is such that the equipotentials are no longer hyperboloidal the field is no longer linear. The resulting equations of ion motion therefore are nonlinear versions of the Mathieu equation for both the x and y trajectory components. They are also cross coupled as can be observed in the following illustrations for the case when circular cylindrical rods are used in place of ideal hyperboloidal cylinders [7]:

$$\ddot{x} + (a - 2q\cos 2\xi)\sum_n C_n \frac{2n+1}{4}\left(\frac{x^2+y^2}{r_0}\right)^{2n} \cdot \{x\cos[2(2n+1)\theta] + y\sin[2(2n+1)\theta]\}.$$

$$\ddot{y} + (a - 2q\cos 2\xi)\sum_n C_n \frac{2n+1}{4}\left(\frac{x^2+y^2}{r_0}\right)^{2n} \cdot \{y\cos[2(2n+1)\theta] - x\sin[2(2n+1)\theta]\}$$

where $\theta = \tan^{-1}\frac{y}{x}$ and the C_n are expansion coefficients. With these nonlinear situations there is little to be gained by modelling the sinusoidal time function since the spatial nonlinearities and cross coupling also present intractabilities. As a result numerical integration methods are still used in these cases for device simulation. This approach to determining the operating characteristics of real instruments has been adopted by Dawson and Whetten [25–28] and by Winters, Musumeci and Richards for a variation on the monopole mass spectrometer [29].

7.2 Dynamic Buckling of Structures

An interesting mechanical situation, in which periodic differential equation descriptions arise, is the buckling of structures brought about by a force that is applied in a periodic manner. To illustrate this consider a thin vertical column, hinged at both ends and end-loaded by a force F as shown in Fig. 7.2.1. If the column has Young's modulus E, second moment of area I and distributed mass per unit length m then the deflection y as a function of position x along the column, and as a function of time, is described by the partial differential equation [30]

$$EI\frac{\partial^4 y}{\partial x^4} + F\frac{\partial^2 y}{\partial x^2} + m\frac{\partial^2 y}{\partial t^2} = 0. \tag{7.2.1}$$

This equation is subject both to boundary conditions in x and initial conditions in time, t. For the moment ignore the latter and observe that at both $x = 0$ and l, where l is the length of the column, $y = 0$ and $\partial^2 y/\partial x^2 = 0$. The last condition arises from the bending moment being zero at the hinges.

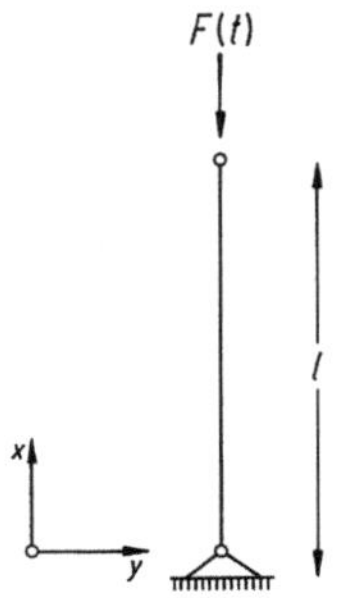

Fig. 7.2.1. Doubly hinged beam subject to end loading

It is assumed that the independent variables are separable and that the solution to (7.2.1) can be expressed as

$$y(x, t) = X(x)f(t). \tag{7.2.2}$$

This will be the case for F constant and is a reasonable approximation when F is a periodic function of time if it is assumed that the period of F is very large in comparison with the period of the fundamental longitudinal vibration of the column and therefore that the axial force in the column can be considered constant with position at any time. Substitution of (7.2.2) into (7.2.1) leads to the pair of independent equations

$$EI\frac{d^4X}{dx^4} + F\frac{d^2X}{dx^2} + a^2X = 0 \tag{7.2.3a}$$

$$m\frac{d^2f}{dt^2} - a^2f = 0 \tag{7.2.3b}$$

where a^2 is a separation constant. A solution to (7.2.3a) is

$$X(x) = A\cos\alpha x + B\sin\beta x.$$

The requirement that $X(x) = 0$ at $x = 0$ demands $A = 0$ whereas $X(x) = 0$ at $x = l$ requires $\beta = n\pi/l$ where n is an integer. Consequently

$$X(x) = B\sin n\pi x/l. \tag{7.2.4}$$

This also satisfies the condition on bending moment at the boundaries so that the solution to (7.2.3a) will not be pursued further. Substitution of (7.2.4) into (7.2.3a) gives

$$a^2 = \left(\frac{n\pi}{l}\right)^2\left[F - EI\left(\frac{n\pi}{l}\right)^2\right]$$

so that (7.2.3b) becomes

$$\frac{d^2f}{dt^2} + \frac{1}{m}\left(\frac{n\pi}{l}\right)^2\left[EI\left(\frac{n\pi}{l}\right)^2 - F\right]f = 0. \tag{7.2.5}$$

For F constant this is an ordinary differential equation the solution of which determines how the deflection of the beam varies with time. It can be seen that the deflection will 'oscillate' (i.e. equation has a sinusoidal solution) if $F < EI(n\pi/l)^2$

whereas for $F > EI(n\pi/l)^2$ the solution contains an increasing exponential term which connotes buckling of the beam. The load at which buckling just commences is called the Euler load for the column. In terms of the first mechanical mode with $n = 1$ this is

$$F_{\text{Euler}} = \pi^2 EI/l^2. \tag{7.2.6}$$

Now suppose $F = P + S\cos\omega t$—i.e. the end loading on the column consists of a constant term and a superimposed periodic force of frequency $\omega\,\text{rad}\cdot\text{s}^{-1}$. Equation (7.2.5) therefore can be expressed, with $n = 1$, as

$$\frac{d^2f}{dt^2} + (A - 2Q\cos\omega t)f = 0 \tag{7.2.7}$$

where

$$A = \frac{\pi^2}{ml^2}\left(\frac{EI\pi^2}{l^2} - P\right)$$

and

$$Q = \frac{S}{2m}\frac{\pi^2}{l^2}.$$

The change of time variable $2\xi = \omega t$ transforms (7.2.7) into the standard Mathieu equation

$$\frac{d^2f}{dt^2} + (a - 2q\cos 2\xi)f = 0$$

with

$$a = \frac{4\pi^2}{m\omega^2 l^2}\left(\frac{EI\pi^2}{l^2} - P\right) \tag{7.2.8a}$$

and

$$q = \frac{2S\pi^2}{m\omega^2 l^2}. \tag{7.2.8b}$$

It is evident therefore from (7.2.2), describing the complete behaviour of the column, that the deflection for periodic end loading will be of the form of a sinusoid of position as in (7.2.4) (which determines the essential flexing shape) multiplied by a Mathieu function of some kind that describes how the amplitude of the flexing changes with time. The initial displacement and transverse velocity of the column (i.e. the initial conditions) establish the kind of the Mathieu function whereas the parameters of (7.2.8) establish its nature. Provided the values of a and q lie in a stable region of the stability diagram for the Mathieu equation the amplitude of the column's vibration remains bounded and buckling will not occur. However should a and q lie within an unstable zone the amplitude grows exponentially thus leading to buckling. Since unstable zones in the stability diagram approach the a axis at squares of integers as depicted in Fig. 6.3, buckling is particularly likely if

$$\frac{4\pi^2}{m\omega^2 l^2}\left(\frac{EI\pi^2}{l^2} - P\right) = K^2 \tag{7.2.9}$$

where K is an integer. This is the case even with very small values of S. Larger values of S will be required for buckling if losses are taken into account as deduced from an examination of the stability diagram for a damped Mathieu equation, such as that shown in Fig. 6.5.

Putting

$$\omega_0^2 = \pi^4 EI/ml^4 \tag{7.2.10}$$

which is the square of the natural frequency of transverse vibration of the column without loading, (7.2.9) can be rearranged as

$$\omega^2 = 4\left(\frac{\omega_0}{K}\right)^2 (1 - P/P_0)$$

where P_0 is the Euler load of (7.2.6).

With $K = 1$, at which instability is most likely, particularly in the presence of losses, it can be seen that a periodic load of infinitesimal amplitude, with frequency in the vicinity of $2\omega_0$ will cause the column to buckle for $P \ll P_0$—i.e. for a static component of the load well under the Euler limit—indeed, ideally, even for no static component. The practical implications of this analysis are quite clear. Should a structure that is designed to withstand a certain static load be subject to a dynamic component with frequency such that (7.2.9) is satisfied (or nearly so) failure of that structure is likely to occur for static components considerably smaller than the design maximum. Such a siutation has been investigated by Walker [31] with respect to dynamic failure of a crane jib under conditions of periodic end loading brought about by the sudden stopping of a descending load on the load line.

In applications of this type it is clear that the actual solutions of the Mathieu equation are of little significance but rather it is stability that is of prime consideration. In particular, a designer would wish to know the extent to which a structure could be loaded dynamically before failure is likely. In a lossless situation this loading is infinitessimal if the frequency of the dynamic load satisfies (7.2.9). In the case of a damped system however the nature of the corresponding Mathieu equation stability diagram indicates that dynamic loads of certain magnitudes can be tolerated even at the so-called parametric frequencies. In these case it is important to be able to determine those critical dynamic loads from an analysis of the stability of the lossy system. This could be achieved by computing values for the characteristic exponent by Hill determinant techniques, as in Sect. 6.1.4. Alternatively the Mathieu equation could be represented by a tractable counterpart, such as the Meissner equation, and characteristic exponents and thus stability thresholds estimated in that manner.

It is interesting to observe, as pointed out in McLachlan [1], that if $EI = 0$ in (7.2.8a), denoting a 'column' with no stiffness such as a stretched string, the corresponding Mathieu equation is

$$\frac{d^2 f}{d\xi^2} - \frac{4\pi^2}{m\omega^2 l^2}[P + S\cos 2\xi]f = 0$$

which is the equation describing the time variation of the 'deflection' of a stretched string of length l, and distributed mass m per unit length subject to a constant component of tension P and a periodic variation in tension of amplitude S and frequency ω. Again stable and unstable vibrations of the string are possible (to the extent that nonlinearities are unimportant) as observed by Melde, in his parametric experiments (see Sect. 1.1).

7.3 Elliptical Waveguides

Waveguides are tubes of constant cross-section manufactured from conducting material and filled with a dielectric medium; often the dielectric is air and from an ideal point of view it is assumed lossless. Electromagnetic energy introduced into a waveguide will propagate in the dielectric just as electromagnetic radiation will propagate in free space. However the presence of the conducting walls of the waveguide modifies the propagation mechanism and allows the energy to be directed almost as desired. Along with transmission lines such as coaxial cables, radiation in free space and, more recently optical fibres, waveguides are channels by which signals can be conveyed for communications and related purposes. The advantages of waveguide propagation are that it is closed to outside interference, it is a convenient means by which signals with frequencies in the GHz range may be controlled and it can handle large bandwidths. Its disadvantages are that it usually has to be constructed from relatively inflexible material and, because of the onset of an effect known as mode conversion—reconversion distortion, cannot be bent in relatively small radii.

There are two broad classes of applications of waveguides. The first is as feeders between transmitters or receivers and antennas in systems such as radar and microwave radio systems. In these cases waveguides with rectangular cross section are generally employed. The second application is as long distance communication channels. These guides are of circular cross section [32]. It is anticipated that this use of waveguide will be superceded in the near future by single mode optical fibres—thin glass fibres of suitable refractive index profiles that operate at light frequencies. These can be viewed also as waveguides but rather than having perfectly conducting boundaries they have discontinuities in refractive index that lead to a so-called radially evanescent wave travelling longitudinally along the outside of a fibre interface in addition to the waveguide-like propagation mechanism inside [33].

Waveguides of elliptical cross section also find application in practice for feeder purposes and are sometimes used in circular waveguide systems for negotiating bends thereby to an extent obviating some of the difficulties sometimes encountered.

This section on elliptical waveguides demonstrates how periodic differential equations arise because of a choice of coordinate system in which to separate a partial differential equation. This is quite a different situation to other applications in which Hill equations result from a system parameter varying periodically with an independent variable.

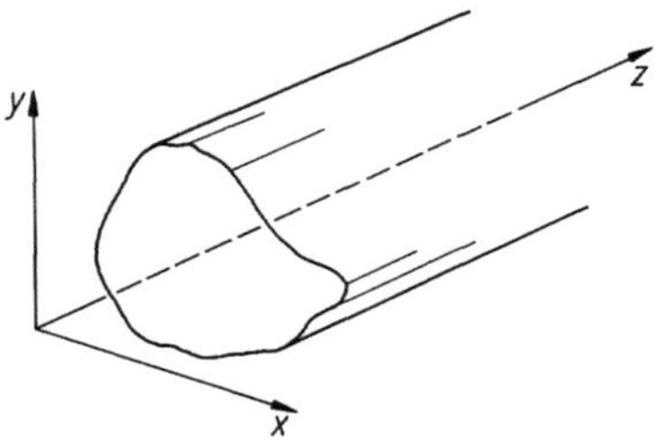

Fig. 7.3.1. Waveguide with constant cross section. The direction of propagation is z

7.3.1 The Helmholtz Equation

Consider a waveguide of constant cross section as shown in Fig. 7.3.1 in which propagation takes place in the z direction. The electric field in such a structure is defined completely by the wave equation

$$\nabla^2 \boldsymbol{E} = \mu\sigma \frac{\partial \boldsymbol{E}}{\partial t} + \mu\varepsilon \frac{\partial^2 \boldsymbol{E}}{\partial t^2} \tag{7.3.1}$$

where μ, σ and ε are the permeability, conductivity and permittivity of the dielectric respectively. ∇^2 is the usual (scalar) Laplacian operator. A similar equation describes the magnetic field. It is assumed that the time dependence of the fields is sinusoidal or can be resolved to sinusoidal form so that

$$\boldsymbol{E} = \underline{E}e^{j\omega t}, \quad \omega = \text{frequency}$$

and the wave equation can be written as

$$\nabla^2 \underline{E} - \gamma_0^2 \underline{E} = 0 \tag{7.3.2}$$

where $\gamma_0^2 = j\omega\mu\sigma - \omega^2\mu\varepsilon$ is the square of the propagation constant for the field if it were unbounded (since boundary conditions have yet to be imposed).

For a lossless dielectric $\sigma = 0$ so that $\gamma_0 = j\omega\sqrt{\mu\varepsilon} \overset{\text{def}}{=} j\beta_0$ giving from (7.3.2)

$$\nabla^2 \underline{E} + \beta_0^2 \underline{E} = 0 \tag{7.3.3}$$

where β_0 is the unbounded phase constant.

It is convenient now to rewrite the three dimensional Laplacian as

$$\nabla^2 = \nabla_t^2 + \frac{\partial^2}{\partial z^2} \tag{7.3.4}$$

where ∇_t^2 is the "transverse" Laplacian—i.e. the operator defined in the two dimensional coordinate system in the cross section of the waveguide, normal to the direction of the propagation.

It is evident that the waveguide as defined is linear and it is known for linear systems (such as free space) that propagation takes place in an exponential manner. Consequently it is of value to assume that the z dependence of $\underline{E}$ is according to $e^{-\gamma z}$ where γ is the propagation constant of the field inside the waveguide. As a result $\partial^2/\partial z^2 = \gamma^2$, which when substituted with (7.3.4) into (7.3.3) leads to the *Helmholtz Equation*

$$(\nabla_t^2 + h^2)\underline{E} = 0 \tag{7.3.5}$$

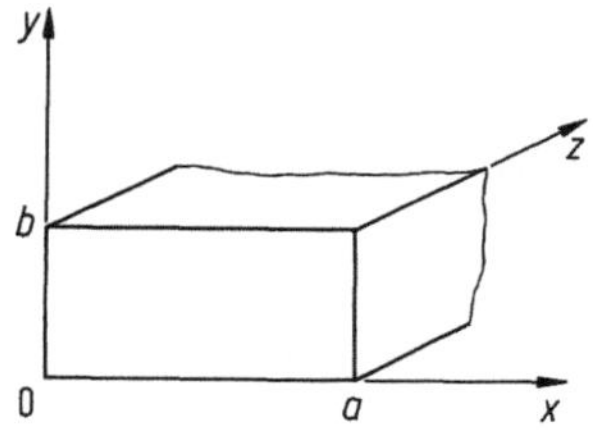

Fig. 7.3.2. Rectangular waveguide cross section showing boundaries and dimensions

with

$$h^2 = \gamma^2 + \beta_0^2 \tag{7.3.6}$$

which is the square of the equation's eigenvalue. The theory of all waveguides of constant cross-section derives from the Helmholtz equation. For a particular cross section a specific form of ∇_t^2 is chosen to facilitate the application of the boundary conditions imposed by the waveguide conductor. The Helmholtz equation is then solved in that coordinate system to yield information on the particular waveguide under consideration. Before proceeding to consider the case of elliptical cross sectional geometry, the case of rectangular and circular waveguides will be considered briefly both to establish some essential theory of a general nature and to highlight the importance of the cross section in determining the type of equations that describe the waveguide field.

7.3.2 Rectangular Waveguides

For a waveguide of any cross section the boundary conditions to be observed by the field are that tangential components of electric field at the conductor must be zero. In the case of the rectangular guide shown in Fig. 7.3.2 these are that

$$E_x = 0 \text{ at } y = 0, b$$

$$E_y = 0 \text{ at } x = 0, a.$$

A convenient coordinate system to use for this structure, that leads readily to separation of the wave equation, is the cartesian coordinate system shown. In this the Helmholtz equation is

$$\left(\frac{\partial^2}{\partial x^2} + \frac{\partial^2}{\partial y^2} + h^2\right) \underline{E} = 0.$$

It is possible to treat the vector equation in $\underline{E}$ as three separate equations. Consequently the component in the direction of propagation can be chosen since the transverse components can be related readily to it via an application of the Maxwell curl equations [34]. It is not important for this treatment to consider the transverse fields. Indeed they are generally of little importance unless sketches of the cross section fields are necessary, such as in the design of waveguide components (attenuators, filters, frequency meters, etc) and in the design of waveguide antennas. Attention is now focussed on a solution to the scalar equation

$$\left(\frac{\partial^2}{\partial x^2} + \frac{\partial^2}{\partial y^2} + h^2\right) E_z = 0. \tag{7.3.7}$$

The z and t dependences have already been taken care of in the terms $e^{j\omega t}e^{-\gamma z} = e^{j\omega t - \gamma z}$. It is sufficient therefore to regard E_z as a function simply of x and y, and to seek a solution of (7.3.7) by separating variables. Thus by putting

$$E_z(x, y) = X(x)Y(y)$$

into (7.3.7) the pair of ordinary differential equations result:

$$\frac{d^2X}{dx^2} + h_x^2 X = 0 \tag{7.3.8a}$$

$$\frac{d^2Y}{dy^2} + h_y^2 Y = 0 \tag{7.3.8b}$$

with

$$h_x^2 + h_y^2 = h^2 \tag{7.3.8c}$$

whereupon solving (7.3.8a) and (7.3.8b) subject to the boundary conditions above leads to the following expression for the longitudinal E field in a rectangular waveguide of dimensions $a \times b$:

$$E_z(x, y, z, t) = E_0 \sin\frac{m\pi x}{a} \sin\frac{n\pi y}{b} e^{j\omega t - \gamma z} \tag{7.3.9a}$$

where m and n are arbitrary integers referred to as mode numbers.

Similarly the longitudinal H field is

$$H_z(x, y, z, t) = H_0 \cos\frac{m\pi x}{a} \cos\frac{n\pi y}{b} e^{j\omega t - \gamma z}. \tag{7.3.9b}$$

It is clear that the use of a rectangular coordinate system has led to a field description in terms of trigonometric functions. This is a result of the linear constant coefficient forms of (7.3.8a) and (7.3.8b). In determining (7.3.9a) it is also seen that

$$h^2 = \left(\frac{m\pi}{a}\right)^2 + \left(\frac{n\pi}{b}\right)^2 \tag{7.3.10}$$

so that the propagation constant for the field is, from (7.3.6)

$$\gamma^2 \stackrel{\text{def}}{=} \gamma_{mn}^2 = h^2 - \beta_0^2 \tag{7.3.11a}$$

$$= \left(\frac{m\pi}{a}\right)^2 + \left(\frac{n\pi}{b}\right)^2 - \omega^2 \mu\varepsilon. \tag{7.3.11b}$$

If $\omega^2\mu\varepsilon > h^2$ the propagation constant is purely imaginary and the field travels without loss in the waveguide. On the other hand if $\omega^2\mu\varepsilon < h^2$ the field decays exponentially with distance travelled. Thus the waveguide exhibits properties of a high pass filter and it is necessary to ensure that the frequency at which it is to be used exceeds the "cut-off" frequency

$$\omega_c = h/\sqrt{\mu\varepsilon}. \tag{7.3.11c}$$

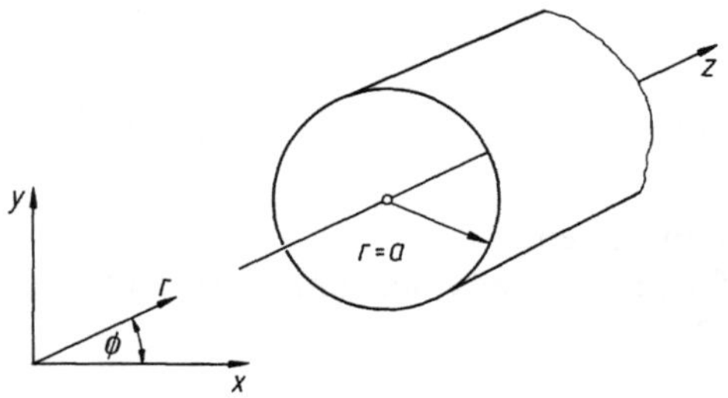

Fig. 7.3.3. Circular waveguide of radius a

The transverse field components are determined entirely by E_z and H_z, so that the whole field in a waveguide is defined in terms of the nature of the longitudinal components. If $E_z = 0$ so that there is only a magnetic field component in the propagation direction the waveguide is said to be carrying an *H mode* whereas if H_z is zero with E_z non zero the waveguide is said to be supporting an *E mode*.[1] Depending upon frequency, a particular waveguide can support several E and H modes simultaneously as determined by the values of the mode numbers m and n in (7.3.10) and (7.3.11). Each mode is generally distinguished by appending the values of m and n as subscripts. Thus there are H_{mn} and E_{mn} modes. The H_{10} mode for example has a cut off frequency given, from (7.3.11), as $\pi/a\sqrt{\mu\varepsilon}$.

7.3.3 Circular Waveguides

The boundary conditions for the circular waveguide of radius a shown in Fig. 7.3.3 are again that the tangential electric field components be zero on the inside surface of the conductor. To make this condition easy to apply and to facilitate analysis of the structure a polar coordinate system is chosen to describe the transverse plane, as shown. Consequently the boundary conditions are

$$E_z, E_\phi = 0 \quad \text{at} \quad r = a.$$

There is also a condition of circular symmetry that has to be obeyed by the field components, viz. that they should remain unchanged under the rotation $\phi = \phi + 2\pi m$, where m is an integer.

The transverse Laplacian in the polar coordinate system is

$$\begin{aligned}\nabla_t^2 &= \frac{1}{r}\frac{\partial}{\partial r}\left(r\frac{\partial}{\partial r}\right) + \frac{1}{r^2}\frac{\partial^2}{\partial\phi^2} \\ &= \frac{\partial^2}{\partial r^2} + \frac{1}{r}\frac{\partial}{\partial r} + \frac{1}{r^2}\frac{\partial^2}{\partial\phi^2}\end{aligned}$$

[1] Alternatively E modes are referred to as transverse magnetic (TM) modes, and H modes as transverse electric (TE) modes.

so that (7.3.5) becomes

$$\left(\frac{\partial^2}{\partial r^2} + \frac{1}{r}\frac{\partial}{r} + \frac{1}{r^2}\frac{\partial^2}{\partial \phi^2} + h^2\right)E_z = 0 \tag{7.3.12}$$

if only the longitudinal E component is considered.

As before the z and t dependences of E_z are already accounted for and E_z is considered just as a function of r and ϕ. Thus let

$$E_z(r, \phi) = R(r)\Phi(\phi)$$

which when substituted in (7.3.12) leads to the separated equations

$$\frac{d^2R}{dr^2} + \frac{1}{r}\frac{dR}{dr} + \left(h^2 - \frac{v^2}{r^2}\right)R = 0 \tag{7.3.13a}$$

and

$$\frac{d^2\Phi}{d\phi^2} + v^2\Phi = 0 \tag{7.3.13b}$$

where v^2 is a separation constant. Equation (7.3.13b) has the solution

$$\Phi(\phi) = A\cos v\phi + B\sin v\phi \tag{7.3.14a}$$

whereas (7.3.13a), which is a Bessel equation, has the solution

$$R(r) = CJ_v(hr) + DY_v(hr). \tag{7.3.14b}$$

Note that $Y_v \to \infty$ as $r \to 0$; thus the second term in (7.3.14b) is not acceptable. Also since the field must remain unchanged if $\phi = \phi + 2\pi m$, v in (7.3.14a) must be an integer—say m. In addition it is not necessary to retain both the sine and cosine terms in (7.3.14a) since the origin of the ϕ coordinate is unspecified and thus can be chosen arbitrarilly. Therefore retain only the cosine term to give as the longitudinal field expression in a circular guide

$$E_z(r, \phi) = E_0 \cos m\phi J_m(hr).$$

Now observing the boundary condition that $E_z = 0$ at $r = a$ requires

$$ha = X^e_{mn}$$

where X^e_{mn} is the nth zero of the mth order Bessel function of the first kind. Thus

$$h = X^e_{mn}/a$$

to give

$$E_z(r, \phi, z, t) = E_0 \cos m\phi J_m\left(X^e_{mn}\frac{r}{a}\right)e^{j\omega t - \gamma z}.$$

In addition (7.3.11c) shows that the cut-off frequency for an E mode in a circular waveguide is

$$\omega_c = X^e_{mn}/a\sqrt{\mu\varepsilon}. \tag{7.3.15a}$$

A similar analysis for H modes, in which the boundary condition is expressed as

$$\frac{\partial H_z}{\partial r} = 0 \quad \text{at} \quad r = a,$$

shows that

$$\omega_c = X^h_{mn}/a\sqrt{\mu\varepsilon} \tag{7.3.15b}$$

for H modes, where X^h_{mn} is the nth zero of the first derivative of the mth order Bessel function.

7.3.4 Elliptical Waveguides

Consider the waveguide of elliptical cross section shown in Fig. 7.3.4 where the interfocal distance of the ellipse is $2f$. Owing to the elliptical geometry and the need to apply an elliptical boundary condition an appropriate coordinate system to adopt is the elliptical system described by orthogonal intersecting ellipses and hyperbolas shown in Fig. 7.3.5. The parameters of this coordinate system are η

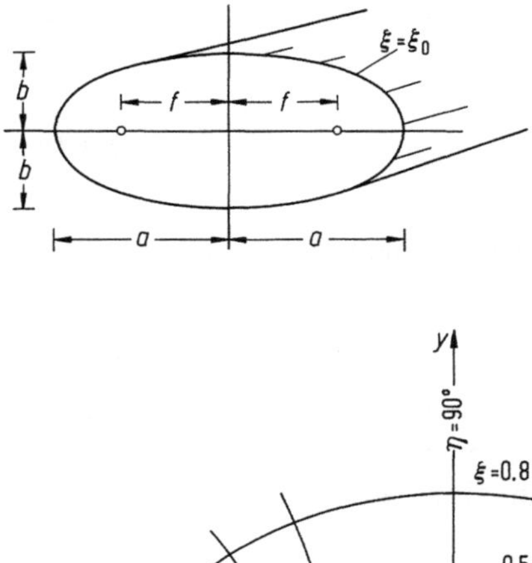

Fig. 7.3.4. Cross section of an elliptic waveguide

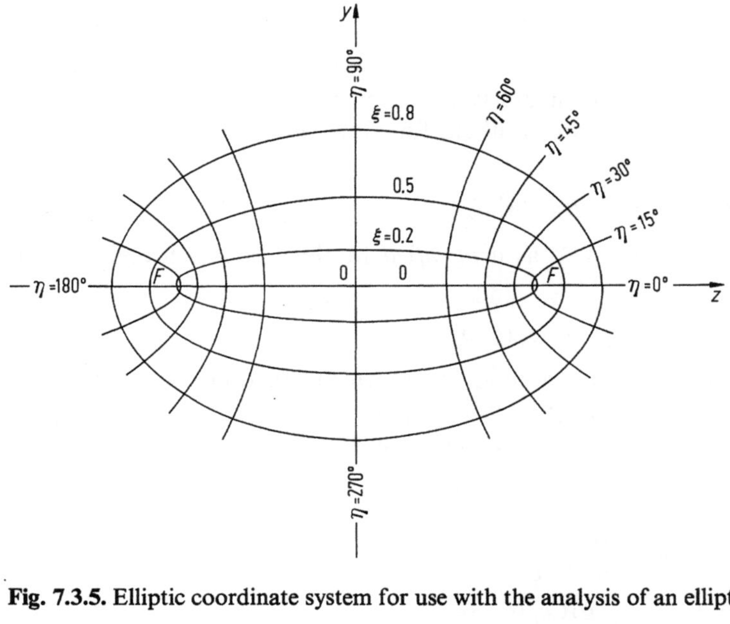

Fig. 7.3.5. Elliptic coordinate system for use with the analysis of an elliptic waveguide

and ξ as shown and in terms of which the transverse Laplacian is from p. 173 of McLachlan [1],

$$\nabla_t^2 = \frac{2}{f^2(\cosh 2\xi - \cos 2\eta)} \frac{\partial^2}{\partial \xi^2} + \frac{\partial^2}{\partial \eta^2}. \tag{7.3.16}$$

Therefore the Helmholtz equation (7.3.5) for the longitudinal electric field becomes

$$\frac{\partial^2 E_z}{\partial \xi^2} + \frac{\partial^2 E_z}{\partial \eta^2} + \frac{h^2 f^2}{2}(\cosh 2\xi - \cos 2\eta) E_z = 0 \tag{7.3.17}$$

A solution is assumed of the form

$$E_z(\xi, \eta) = X(\xi) N(\eta)$$

which when substituted in (7.3.17) yields the separated equations

$$\frac{\mathrm{d}^2 N}{\mathrm{d}\eta^2} + (a - 2q \cos 2\eta) N = 0 \tag{7.3.18a}$$

and

$$\frac{\mathrm{d}^2 X}{\mathrm{d}\xi^2} - (a + 2q \cosh 2\xi) X = 0 \tag{7.3.18b}$$

with

$$q = (hf/2)^2 \tag{7.3.18c}$$

and a is a separation constant.

Equation (7.3.18a) will be recognised as the Mathieu equation for which a general solution is, from (6.8),

$$N(\eta) = Ace_\nu(\eta, q) + Bse_\nu(\eta, q).$$

Now $N(\eta)$ must be periodic with η as η changes by multiples of 2π, as is evident from the definition of the coordinate system shown in Fig. 7.3.5. As a result ν must be an integer, say m, since Mathieu functions which are periodic are of integral order. Thus

$$N(\eta) = Ace_m(\eta, q) + Bse_m(\eta, q). \tag{7.3.19a}$$

Equation (7.3.18b) is referred to as a *modified* Mathieu equation and has the general solution

$$X(\xi) = CCe_m(\xi, q) + DSe_m(\xi, q) \tag{7.3.19b}$$

where $Ce_m(\xi, q)$ and $Se_m(\xi, q)$ are modified Mathieu functions of the first and second kind, respectively, of integral order m [1].

From (7.3.19a) and (7.3.19b) it appears that the longitudinal E field is

$$E_z(\xi, \eta) = (Ace_m(\eta, q) + Bse_m(\eta, q))(CCe_m(\xi, q) + DSe_m(\xi, q)). \tag{7.3.20}$$

However to ensure continuity of E_z and $\mathrm{d}E_z/\mathrm{d}\xi$ with η it can be shown that the only allowed terms in (7.3.20) are [1]

$$E_z(\xi, \eta) = E_{oc} ce_m(\eta, q) Ce_m(\xi, q) \tag{7.3.21a}$$

and

$$E_z(\xi, \eta) = E_{os} se_m(\eta, q) Se_m(\xi, q). \tag{7.3.21b}$$

It is usual to distinguish these as two separate E modes in an elliptical waveguide. They arise because of the eccentric nature of the guide's cross section which gives it two axes of symmetry, as compared with the completely circular symmetric character of the circular waveguide. A similar situation exists for H modes.

The field E_z must be zero at the inside surface of the guide conductor. If this surface is located on the ellipse given by $\xi = \xi_0$ then (from (7.3.21 a or b)

$$Ce_m(\xi_0, q) = 0$$

and

$$Se_m(\xi_0, q) = 0.$$

Thus the only values that can be taken on by q are those that give zeros in the modified Mathieu functions for $\xi = \xi_0$. Let these values of q be denoted q^c_{mn} and q^s_{mn}, interpreted as the nth values of q that give zeros of the mth order modified Mathieu functions of the first and second kinds respectively. Consequently from (7.3.18c) the eigenvalues, h, for elliptical waveguide E modes are

$$h = \frac{2}{f}\sqrt{q^{c \text{ or } s}_{mn}}$$

giving cut-off frequencies of

$$\omega_c = 2\sqrt{q^{c \text{ or } s}_{mn}}/f\sqrt{\mu\varepsilon}. \tag{7.3.22a}$$

In a similar manner the cut off frequencies for H modes are given by

$$\omega_c = 2\sqrt{q'^{c \text{ or } s}_{mn}}/f\sqrt{\mu\varepsilon} \tag{7.3.22b}$$

where q'^c_{mn} and q'^s_{mn} are the nth values of q that give zeros in the first derivative with ξ of the modified Mathieu functions of order m.

An extension of the theory of simple elliptical waveguides is that of elliptical guides containing a confocal partial dielectric or lined with a dielectric with confocal elliptical boundaries. The confocal nature of the dielectric boundary renders the field equations again separable in an elliptic co-ordinate system leading to Mathieu equation descriptions. The modes can then again be expressed in Mathieu functions and are hybrid in nature as discussed by Rengarajan and Lewis [34a]. In a similar manner Rengarajan and Lewis [34b] have treated the surface wave transmission line with an elliptical cross section. Their study relates particularly to an elliptical conductor (which they refer to as an elliptical Goubau line) coated with an elliptically confocal dielectric.

Yamashita, Atsuki and Nishino [34c] have considered wave propagation along composite elliptic dielectric waveguides. Such guides are considered of value for single polarisation propagation in optical fibres and consist of an elliptical core surrounded by an elliptical cladding that is not necessarily confocal with the core. As a result an elliptic cylinder co-ordinate system is of little value for separating the wave equation. Instead Yamashita et al use expansions of circular field components described by Bessel functions.

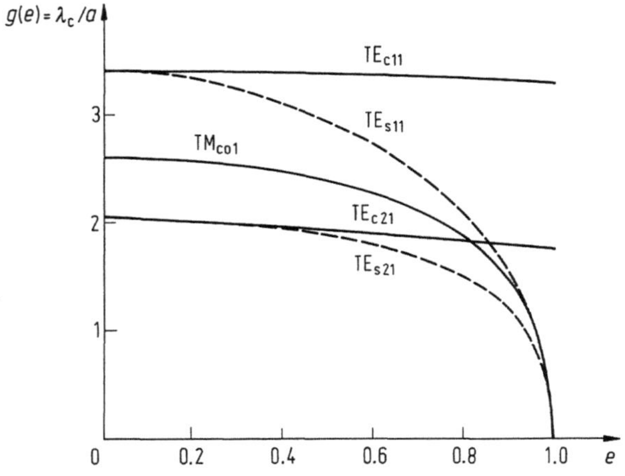

Fig. 7.3.6. The function $g(e) = \lambda c/a$ versus e for five consecutive modes in an elliptic waveguide; [35]

7.3.5 Computation of the Cut-off Frequencies for an Elliptical Waveguide

A difficulty with trying to calculate the cut-off frequencies of an elliptical waveguide is the need to know the Mathieu function zeros q_{mn}^c, q_{mn}^s, $q_{mn}'^c$, $q_{mn}'^s$ in (7.3.22a) and (7.3.22b). This problem has been addressed by Kretzschmar [35] who used Bessel function product series to represent the necessary Mathieu functions. His results can be used if (7.3.22a) and (7.3.22b) are modified slightly.

If e is the eccentricity of the elliptical cross section of the guide and its major and minor dimensions are $2a$ and $2b$ as depicted in Fig. 7.3.4, then both (7.3.22a) and (7.3.22b) can be expressed

$$\omega_c = 2\sqrt{q}/ae\sqrt{\mu\varepsilon}$$

where q is q_{mn}^c etc., as appropriate and c is the velocity of light in the waveguide dielectric. This can be recast in terms of cut-off wavelength λ_c $(= 2\pi/\omega_c\sqrt{\mu\varepsilon})$ as

$$\lambda_c = \pi ae/\sqrt{q}.$$

Thus

$$\frac{\lambda_c}{a} = \pi e/\sqrt{q} = g(e) \tag{7.3.23}$$

showing that the ratio of cut-off wavelength for a given mode, to the semi-major dimension of the guide is a function only of the eccentricity of the cross-section. Kretzschmar has computed the function $g(e)$ for the first 19 modes of an elliptical waveguide for eccentricities between 0.0 and 0.95. Some modes are given in Fig. 7.3.6 which shows that the dominant mode of an elliptical guide is the H_{c11} mode. (The dominant mode is that with the lowest cut-off frequency or highest cut-off wavelength). For an eccentricity of zero this degenerates (with the H_{s11} mode) to the H_{11} dominant mode of a circular waveguide. Kretzschmar has also given approximate, but useful formulas for determining the value of q for the eight lowest order modes, as a function of eccentricity. For the H_{c11} mode in particular q is given by

$$q \simeq \quad 0.8476e^2 - 0.0013e^3 + 0.0379e^4 \qquad\qquad 0 \leq e < 0.4$$

$$\simeq -0.0064e + 0.8838e^2 - 0.0696e^3 + 0.0820e^4 \quad 0.4 \leq e \leq 1.0.$$

As an illustration of the use of these results consider the calculation of the dominant cut-off frequency of an elliptical guide with dimensions $a = 20$ mm, $b = 10$ mm. For comparison the dominant cut off frequencies of a 40 mm × 20 mm rectangular guide (H_{10}) and a 40 mm diameter circular guide (H_{11}) will also be found.

For the elliptic guide $e = 0.866$ so that

$$q_{mn}^c \simeq 0.6582.$$

This gives a cut-off wavelength of 67.1 mm and thus a cut-off frequency of 4.47 GHz if the guide is air-filled ($\varepsilon = \varepsilon_0 = 8.85$ pF m^{-1}, $\mu = \mu_0 = 400\pi$ nH m^{-1}). From (7.3.11c) and (7.3.10) it can be established that the cut-off frequency of the rectangular guide is 3.75 GHz, corresponding to a cut-off wavelength of 80.0 mm, whilst (7.3.15b), along with a table of Bessel function zeros [32] shows that the cut-off frequency and wavelength of the circular guide are 4.40 GHz and 68.3 mm respectively.

7.4 Wave Propagation in Periodic Media

An application of periodic differential equations, quite different to those treated previously, is the propagation of waves through media that have parameters that vary periodically in the direction of propagation. Whereas the previous illustrations have involved only one independent variable, wave propagation in periodic media involves the behaviour of waves with *time*, as they 'travel' *through* a structure whose properties vary with *position*. It is the interaction of the time nature of the wave and the spatial nature of the medium that makes this situation unique.

The essential points of this application will be developed using the general form of the lossless wave equation. If the medium in which the wave is propagating is considered infinite in the x and y directions, and propagation takes place in the z direction, then the equation is

$$\frac{\partial^2 \psi}{\partial z^2} - \frac{1}{v^2}\frac{\partial^2 \psi}{\partial t^2} = 0 \tag{7.4.1}$$

where v is the phase velocity of the wave and ψ is the wave under consideration. In the case of an electric transmission line ψ could be voltage and $c = (lc)^{-1/2}$ where l and c are the distributed inductance and capacitance per unit length of the line respectively. Alternatively, for electromagnetic wave propagation in free space ψ could be the electric or magnetic field vectors whereas $v = (\mu\varepsilon)^{-1/2}$ where μ and ε are medium permeability and permittivity. The equation could just as well describe acoustic waves.

The treatment to be given here only seeks to consider the special aspects of the theory of periodic systems that are relevant to applications of the wave propagation type. It is not possible to be comprehensive since each particular application involves extensive knowledge of the relevant fields. Nevertheless the material to be determined is common to most manifestations of periodic structures. A detailed account of early work in the field will be found in the classic treatment of Brillouin [36]. Whilst it introduces most of the concepts used to describe this aspect of wave propagation it is, from an applications point of view, a little dated. Moreover much of its approach is not based explicitly upon the theory of periodic differential equations. A more up-to-date and comprehensive review of applications—with reference to propagation in both active and passive periodic structures—has been given by Elachi [37]. Elachi also discusses the case of wave propagation with periodic boundary conditions. The particular case of wave scattering from periodic surfaces has been reviewed by Chuang and Kong [37a].

7.4.1 Pass and Stop Bands

Assume that $\psi(z, t)$ in (7.4.1) varies sinusoidally with time so that

$$\psi(z, t) = \Psi(z)e^{j\omega t}.$$

Substitution of this form into (7.4.1) gives

$$\frac{d^2\Psi}{dz^2} + \frac{\omega^2}{v^2}\Psi = 0. \tag{7.4.2}$$

Suppose now that a parameter of the medium through which the wave is propagating varies periodically with position such that the phase velocity changes according to

$$\frac{1}{v(z)^2} = s_0 - s_1 g(z), \quad g(z) = g(z + \theta) \tag{7.4.3}$$

where $g(z)$ has no average value and has unit amplitude.

A substitution such as that in (7.4.3) cannot always be done simply and, indeed, for each particular application the full wave equation should be derived from basic principles to ensure that terms are not missing. A case in point is the magnetic field vector in a medium with varying permittivity as demonstrated in Sect. 7.4.3 following. Further, a return to the basic Telegrapher's equations shows, for a transmission line with z dependent inductance and capacitance, that the equations to be considered are

$$\frac{d^2V}{dz^2} - \frac{1}{l(z)}\frac{dl(z)}{dz}\frac{dV}{dz} + \omega^2 l(z)c(z)V = 0$$

and

$$\frac{d^2I}{dz^2} - \frac{1}{c(z)}\frac{dc(z)}{dz}\frac{dI}{dz} + \omega^2 l(z)c(z)I = 0.$$

If the distributed inductance is constant the voltage equation is of the form of (7.4.2) but the current equation is not. Rather it contains a first derivative that has to be removed (for specific $c(z)$ variations) by an application of (2.4).

Substitution of (7.4.3) into (7.4.2) gives

$$\frac{d^2\Psi}{dz^2} + \omega^2(s_0 - s_1 g(z))\Psi = 0.$$

It is expedient to introduce the change of independent variable

$$\zeta = \pi z/\theta$$

whereupon the wave equation becomes

$$\frac{d^2\Psi}{d\zeta^2} + (a - 2qg(\zeta))\Psi = 0 \tag{7.4.4}$$

where $g(\zeta)$ has a period of π and

$$a = \theta^2\omega^2 s_0/\pi^2 \tag{7.4.5a}$$

$$q = \theta^2\omega^2 s_1/2\pi^2. \tag{7.4.5b}$$

Equation (7.4.4) is in the canonical form of a Hill equation so that well-known properties can now be exploited to discover the nature of wave propagation in periodic media. For example, it is known from Floquet theory that the function $\Psi(\zeta)$ can be expressed

$$\Psi(\zeta) = e^{\pm\mu\zeta} \sum_{r=-\infty}^{\infty} C_r e^{j2r\zeta} \tag{7.4.6}$$

as noted in Sects. 2.4 and 4.3.6. If the characteristic exponent μ is real or complex the wave function $\Psi(\zeta)$ will decay with propagation (both in the positive and negative ζ directions). An exponentially increasing solution must be precluded by the fact that energy cannot be supplied to the wave if the medium through which it travels is passive. If μ is purely imaginary $\Psi(\zeta)$ will obviously be a pair of wave functions travelling in opposite directions, together given by

$$\begin{aligned}\Psi(\zeta) &= e^{\pm j\beta\zeta} \sum_{r=-\infty}^{\infty} C_r e^{j2r\zeta} \\ &= \sum_{r=-\infty}^{\infty} C_r e^{j(\pm\beta+2r)\zeta} \\ &= \sum_{r=-\infty}^{\infty} C_r e^{\pm j\beta_r\zeta}.\end{aligned} \tag{7.4.7}$$

Equation (7.4.7) shows that $\Psi(\zeta)$ is made up of sets of harmonics—called *space harmonics* in the context of the application—that travel through the medium, potentially in both directions. The wave numbers of the individual harmonics that the function is resolvable into are given by

$$\beta_r = \beta + 2r. \quad -\infty \le r \le \infty. \tag{7.4.8}$$

Consequently the nature (and value) of the characteristic exponent (β or μ) of the particular Hill equation is important in determining the nature of the propagation. Since the characteristic exponent is summarised in the stability diagram for the equation it is of value at this stage to relate the situation at hand to such a diagram.

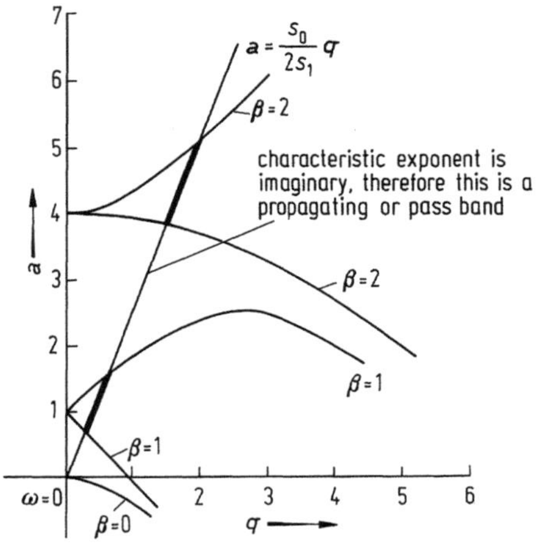

Fig. 7.4.1. Stability diagram showing how pass and stop bands in frequency arise for wave propagation in a periodic medium

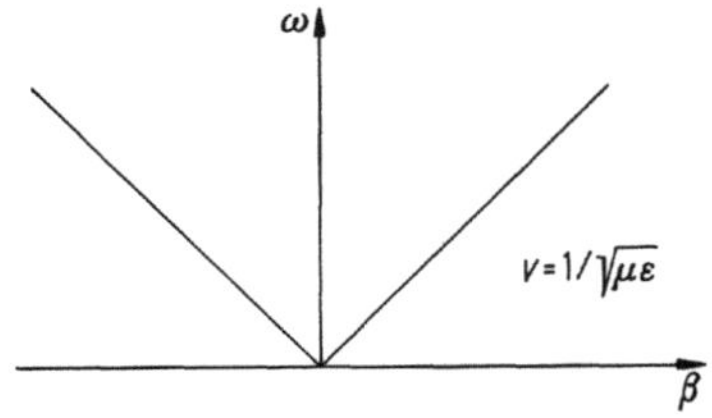

Fig. 7.4.2. Dispersion curve for lossless, free space propagation. Note that the phase velocity, given by ω/β, is constant

Suppose the stability diagram for (7.4.4) is that shown in Fig. 7.4.1. Consider now how the characteristic exponent for $\Psi(\zeta)$ can be determined, in principle, from the diagram in view of the definitions of a and q given in (7.4.5). For a given medium the values of s_0, s_1 and θ are fixed whereas it is possible to envisage ω as variable—e.g. by reason of how the 'field' $\Psi(\zeta)$ is generated. Both a and q are quadratically dependent upon ω. They are both zero for $\omega = 0$ and increase together as ω increases, so that their ratio stays constant. A complete picture of how the periodic medium appears to waves of different frequencies can be obtained therefore by plotting a line of slope

$$\frac{a}{q} = \frac{s_0}{2s_1}$$

on the stability diagram. This is shown in Fig. 7.4.1. From an examination of this equation it is clear that for some frequency bands the characteristic exponent is imaginary, and thus waves of those frequencies will propagate in the medium, whereas for other ranges of frequency the characteristic exponent is real (strictly complex) showing that the wave does not propagate and in fact is evanescent. The former frequency bands are called *pass* bands whereas the latter are referred to as *stop* bands.

7.4.2 The $\omega - \beta_r$ (Brillouin) Diagram

It is customary in wave propagation problems to summarise the nature of the propagation mechanism in an $\omega - \beta$ dispersion curve, where β is the phase constant of the propagating wave. For example, in the case of free space electromagnetic wave propagation in a lossless medium it can be shown that [34] $\beta^2 = \omega^2\mu\varepsilon$ so that $\omega = \pm\beta/\sqrt{\mu\varepsilon}$. Consequently the dispersion curve is linear as shown

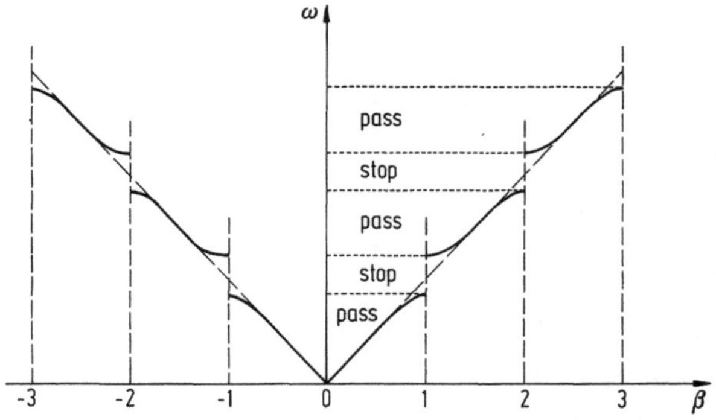

Fig. 7.4.3. ω as a function of β derived from the line of slope $s_0/2s_1$ in Fig. 7.4.1. This is the first stage in the $\omega - \beta_r$ diagram for a periodic medium

in Fig. 7.4.2, indicating that waves of different frequency travel with the same phase velocity (given by $v = \omega/\beta$)—i.e. they do not disperse with time.

The $\omega - \beta$ diagram for propagation in a periodic medium can be derived from (7.4.8) if β as a function of frequency can be determined explicitly. Construction of such a diagram will be approached here in two stages. First the frequency dependence of the characteristic exponent will be determined in general terms. Next the additive $2r$ term in (7.4.8) will be accounted for. To address the former it is necessary to consider the value of β in some detail.

From (2.27b) it can be seen that the characteristic exponent on the stability boundaries in Fig. 7.4.1 is given by

$$\mu = \mathrm{j}m \quad m = 0, \pm 1, \text{etc.}$$

since trace $\{\phi(\pi, 0)\} = \pm 2$ and $\theta = \pi$. Thus $\beta = m$ on the edges of pass bands. Furthermore since $\Delta(0) = 1$ for $q \to 0$, (4.25) shows that $\beta = \pm\sqrt{a}$ on the a axis. Stability boundaries emanate from $a = m^2$ where m is an interger, so that the pair of boundaries that start at $a = 1$ each correspond to $\beta = \pm 1$. Those from $a = 4$ correspond to $\beta = \pm 2$ and so on. The boundaries therefore correspond to the integral values of β shown in Fig. 7.4.1. Consequently, within a region of stability β is bounded as

$$|m| \leq \beta \leq |m + 1|. \tag{7.4.9}$$

Its explicit variation in a particular region would have to be determined by means of suitable analytical tools.

The range of frequencies supported by a particular pass band can be determined by noting the values of a at the band edges and making use of (7.4.5a). If the characteristic exponent has also been determined as a function of a (between the bounds given by 7.4.9) the diagram shown in Fig. 7.4.3 can be constructed, in which β is shown as a linear abcissa scale.

It is evident that as $q \to 0$ (corresponding to $s_1 \to 0$ in (7.4.3), so that there is no periodic perturbation of the medium properties), the $\omega - \beta$ diagram of Fig. 7.4.3 is described exactly by $\beta = \pm\sqrt{a}$, which from (7.4.5a) gives

$$\beta = \pm\theta\frac{\sqrt{s_0}}{\pi}\omega.$$

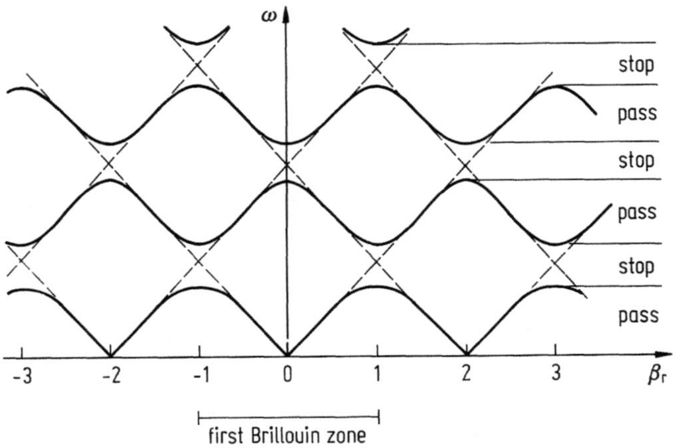

Fig. 7.4.4. Brillouin diagram (complete dispersion curve) for wave propagation in a periodic medium

Here β is the phase constant when ζ is the independent variable. The phase constant in the actual medium is $\beta\pi/\theta$ as can be inferred from (7.4.7) and the change of variable $\zeta = \pi z/\theta$. Consequently the unperturbed phase constant is

$$\beta = \pm\sqrt{s_0}\omega = \omega/v,$$

which is the well known $\omega - \beta$ relation for free space propagation in an unperturbed medium. As q (i.e. s_1) increases from zero the stop bands illustrated in Fig. 7.4.3 develop at integral values of β, as shown. They grow as perturbations from the free space, constant parameter $\omega - \beta$ lines.

The $\omega - \beta_r$ diagram can now be considered in detail using (7.4.8). That expression shows that the values of β_r for the various space harmonics are simply the value of β shown in Fig. 7.4.3 but offset by $2r$ with $r = \pm 1, \pm 2$ etc. Traversing a complete range of r in each pass band simply repeats the curve of Fig. 7.4.3 to give the so-called Brillouin diagram of Fig. 7.4.4. Owing to the periodicity with β of the curves in Fig. 7.4.4, Brillouin [36] suggests that only the first region, given by $-1 \leq \beta \leq 1$, be retained. He calls this the first zone although it is now widely known as the first Brillouin zone.

To determine exactly the edges of the pass bands and thus to know exactly the frequencies of propagation that will be supported by a particular periodic medium it is necessary to compute by some means the exact form of the segments in the $\omega - \beta_r$ diagram of Fig. 7.4.3. This would also of course yield dispersion information. This computation is not straightforward and remains a difficult problem although it has been approached by employing truncated Hill determinants [37]. However, a particularly simple means for obtaining a nearly exact $\omega - \beta_r$ curve is to use the modelling technique of Chap. 5. To do this the actual medium perturbation would be replaced by a counterpart for which the discrete transition matrix $\phi(\theta, 0)$ can be determined analytically. Equation (2.27b) can then be used to give β as a function of a, which in turn is related to frequency by (7.4.5a).

7.4.3 Electromagnetic Wave Propagation in Periodic Media

Consider now a particular example of wave propagation, viz. that of the passage of electromagnetic waves through a medium which has a permittivity that varies periodically in the direction of propagation. The appropriate wave equations can be derived from the Maxwell equations which, for a source-free, lossless medium can be expressed

$$\nabla \times \boldsymbol{E} = -\mathrm{j}\omega\mu\boldsymbol{H} \tag{7.4.10a}$$

$$\nabla \times \boldsymbol{H} = \mathrm{j}\omega\varepsilon(z)\boldsymbol{E}, \tag{7.4.10b}$$

in which a sinusoidal time dependence has been assumed. Taking the curl of (7.4.10a) gives

$$\nabla \times \nabla \times \boldsymbol{E} = -\mathrm{j}\omega\mu\nabla \times \boldsymbol{H}$$

whereupon using (7.4.10b) yields

$$\nabla \times \nabla \times \boldsymbol{E} - \omega^2\mu\varepsilon(z)\boldsymbol{E} = 0. \tag{7.4.11}$$

Since $\varepsilon(z)$ is periodic with position (7.4.11) represents a three dimensional Hill equation. An equation for the magnetic field vector is obtained in a similar manner. Taking the curl of (7.4.10b) gives

$$\nabla \times \nabla \times \boldsymbol{H} = \mathrm{j}\omega\nabla \times (\varepsilon(z)\boldsymbol{E})$$

$$= \mathrm{j}\omega\{\varepsilon(z)\nabla \times \boldsymbol{E} + \nabla\varepsilon(z) \times \boldsymbol{E}\}.$$

Substitution from (7.4.10a) leads to

$$\nabla \times \nabla \times \boldsymbol{H} - \frac{\nabla\varepsilon(z)}{\varepsilon(z)}\nabla \times \boldsymbol{H} - \omega^2\mu\varepsilon(z)\boldsymbol{H} = 0 \tag{7.4.12}$$

Again for $\varepsilon(z)$ periodic the $\boldsymbol{H}$ vector is given by the solution to a Hill equation—however in this case with a first order space derivative term.

Since the magnetic and electric field vectors are given by solutions to quite different Hill equations H and E modes will have different stop and pass band structures in the dielectric medium. Tamir et al [38] have considered the details of H mode propagation for which an electric field vector exists in only one transverse coordinate. They show that if the permittivity varies sinusoidally with position the resulting wave equation is of the form of the Mathieu equation. From this they determine the appropriate Brillouin diagrams. Further, they demonstrate that the various space harmonics propagate at different angles to the direction of propagation for a constant permittivity. The case of E modes has been dealt with by Yeh et al [39]. In fact their treatment commences with a generalisation that covers all wave modes. It then concentrates on E modes and, in view of the form of (7.4.12) demonstrates that even when the permittivity varies sinusoidally the wave equation for E modes is quite a general Hill equation. Having defined $\varepsilon(z)$ explicitly they employ (2.4) to arrive at a Hill equation (in terms of a suitable scalar potential function) which is analysed by means of Hill determinants to give stability diagrams and then Brillouin zones.

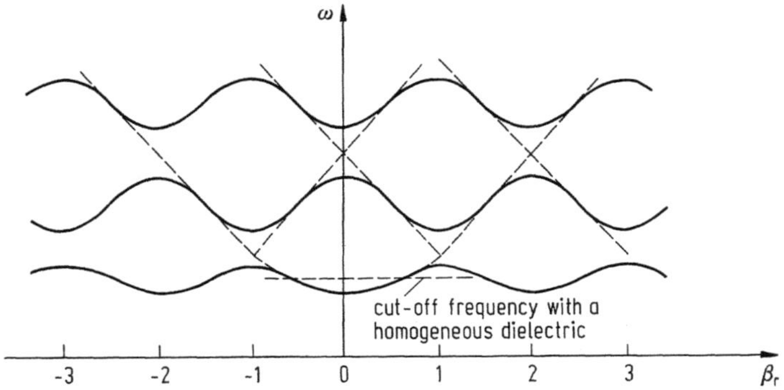

Fig. 7.4.6. Dispersion curve for a waveguide with a periodic dielectric. It is of interest to note that the cut-off frequency is lower than that of a guide with a homogeneous dielectric of the same average value. This is discussed by Tamir et al. [38] and Yeh et al. [39]

where

$$a = 2m\theta^2 E/\hbar^2\pi^2 \tag{7.4.16a}$$

$$q = m\theta^2 V_m/\hbar^2\pi^2 \tag{7.4.16b}$$

in which V_m is the amplitude of the potential $V(z)$, and $v(\zeta)$ has unit amplitude. This distinction is not crucial but may help when comparing (7.4.15) to other versions of the Hill equation. As in Sect. 7.4.1 we express $\psi(\zeta)$ in Floquet form

$$\psi(\zeta) = \mathrm{e}^{\pm \mathrm{j}\beta\zeta} \sum_{r=\infty}^{\infty} C_r \mathrm{e}^{\mathrm{j}2r\zeta}$$

$$= \sum_{r=-\infty}^{\infty} C_r \mathrm{e}^{\pm \mathrm{j}\beta_r z}$$

where $\beta_r = \beta + 2r$, $r = 0, \pm 1$ etc. is the phase constant, or wave number, or wave vector of the rth 'space harmonic' of the electron wave, 'travelling' in a medium of periodicity π. As in the previous discussions the wave vector referred to the lattice periodicity of θ is given by

$$\beta_r = (\beta + 2r)\frac{\pi}{\theta} \tag{7.4.17}$$

in view of the change of independent variable that led to (7.4.15).

Using (7.4.16a) and (7.4.16b) and an assumed stability diagram for (7.4.15) it is possible now to explore the existence or otherwise of electrons with given energy in the crystal lattice. Suppose the stability diagram for (7.4.15) is that illustrated in Fig. 7.4.7. For a given crystal potential, q in (7.4.16b) is a constant whereas a in (7.4.16a) varies directly with electron energy. This is shown as a vertical line in Fig. 7.4.7 depicting that there are certain values of a and thus electron energy that give rise to stable solutions, whereupon the wave function is bounded implying

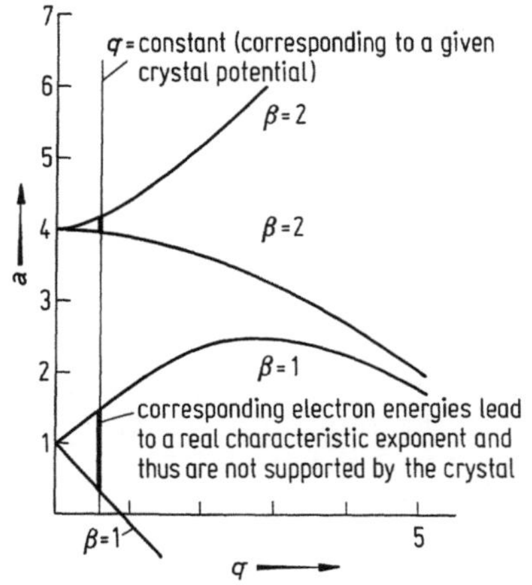

Fig. 7.4.7. Stability diagram for electrons in a single dimensional crystal lattice, illustrating the line of constant lattice potential

a non-zero probability of finding electrons with those energies. Values of a in unstable regions correspond to electron energies that give unstable solutions to the Schrödinger equation. These imply that the wave functions approach zero (in both $\pm\zeta$ directions) and consequently there is very little chance of finding electrons with those energies. The stable regions therefore correspond to the *allowed* energy bands referred to in connection with crystals such as semiconductor compounds whereas the unstable regions of the stability diagram are the *forbidden* bands.

The band structure of a crystal is commonly summarised in an energy—momentum diagram—this is a Brillouin diagram of E vs. p where p is electron momentum. However since $p = \hbar\beta$, where β is the wave number of the electron, the energy—momentum diagram follows directly from an $E - \beta$ diagram analogous to the $\omega - \beta$ diagrams of the previous sections.[2] As before, it can be constructed in the following manner.

In accord with the discussion relating to (7.4.9) the stability boundaries in Fig. 7.4.7 correspond to the β values shown. These values are the ones that will be encountered on the edges between the allowed and forbidden bands. Within the allowed bands β can be related to a (along the a = constant line in Fig. 7.4.7) if the Schrödinger equation in (7.4.15) can be solved analytically. This of course depends upon the form of $v(\zeta)$. In (7.4.16a) we have a relationship between electron energy E and a. By cancelling a as a common factor between these two relations the $E - \beta$ diagram depicted in Fig. 7.4.8 will be obtained. The specific shapes of the individual segments depend upon the actual form of $v(\zeta)$. Since $v(\zeta)$ in a crystal is a complicated function very little analytical prediction of the

[2] In quantum mechanics it is usual to denote wave number by k. The symbol β will be retained here for continuity with other sections.

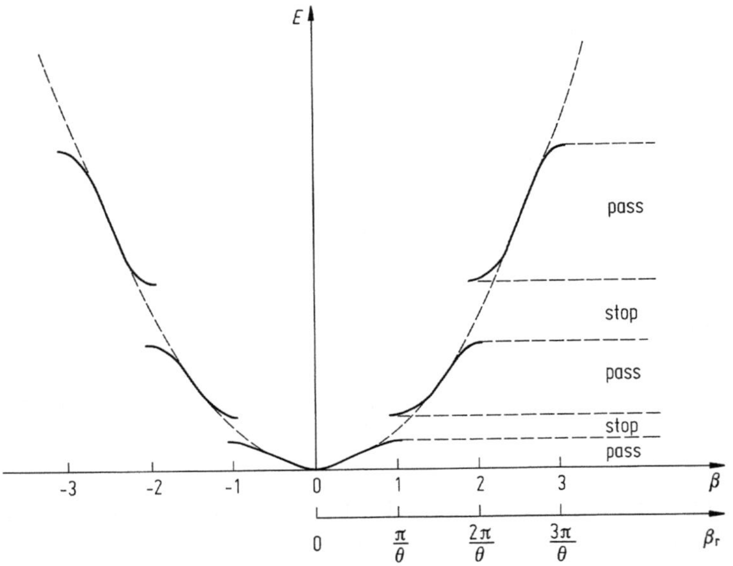

Fig. 7.4.8. Energy as a function of wave number (phase constant) for electrons in a single dimensional crystal. This is the first stage in a complete $E-\beta$ diagram

band structure of a particular crystal is possible. Kronig and Penny [40] in a celebrated paper derived band properties by approximating $v(\zeta)$ by a series of impulses; however if an actual form for $v(\zeta)$ can be deduced or reasonably approximated then it can be represented by a series of steps and ramps and the corresponding $E - \beta$ diagram deduced according to the methods presented here in Chap. 5.

When $V(z) = 0$, i.e. when the varying potential seen by the electron is reduced to zero, $\beta = \sqrt{a}$ as noted in Sect. 7.4.2, which gives

$$\beta = \pm\frac{\theta}{\hbar\pi}\sqrt{2mE}. \tag{7.4.18a}$$

Multiplying β by π/θ to correct its value to that for the real spatial axis z instead of the canonical variable ζ we have

$$\beta = \pm\hbar^{-1}\sqrt{2mE} \tag{7.4.18b}$$

as the dispersion curve for an electron in free space. Since $p = \hbar\beta$ this gives

$$2mE = p^2$$

or

$$E = \frac{p^2}{2m}$$

as is well-known. Further (7.4.18a) and (7.4.18b) each describe the essential parabolic shape of the $E - \beta$ diagram of Fig. 7.4.8.

In view of (7.4.17) it is clear that the basic diagram of Fig. 7.4.8 can be repeated

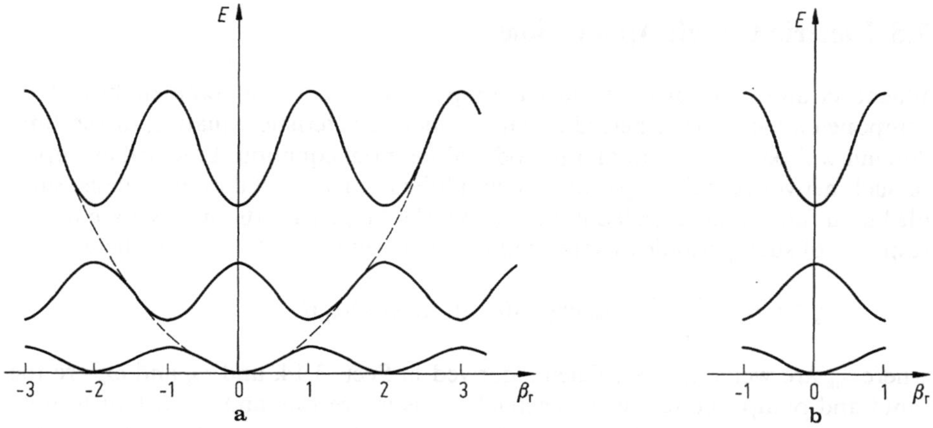

Fig. 7.4.9. a Complete energy-wave number (energy-momentum) diagram for electrons in a single dimensional crystal lattice; **b** the first Brillouin zone of the energymomentum diagram

periodically as shown in Fig. 7.4.9a to account for all possible space harmonics of the electron waves. As before this allows us to concentrate solely upon the region $-1 \leq \beta_r \leq 1$ as an $E - \beta_r$ diagram as shown in Fig. 7.4.9b. Again this is commonly referred to as the *first Brillouin Zone*.

A readable account of the material presented in this section will be found in Ferry and Fannin [41] particularly with regard to the Kronig and Penny model. A more advanced treatment, based substantially upon Floquet theory has been presented by Weinreich [42].

7.4.6 Other Examples of Waves in Periodic Media

Elachi [37] has reviewed a broad cross-section of modern applications of wave propagation through structures and media with periodically distributed parameters. Some of these relate to passive media such as surface acoustic wave electronic devices, optical filters, the propagation of water waves over corrugated sea beds and the compound eyes of insects. Other modern situations concern active media with periodic perturbations such as distributed feedback lasers.

The study of wave-wave interactions for applications such as parametric amplification in travelling wave devices leads to a consideration of space-time periodic media. Analysis of space-time periodicity can be found in the recent papers by Seshadri [42a, 42b]. Seshadri [42c] has also determined reflection and transmission coefficients for waves incident on a sinusoidally periodic dielectric slab.

Recent interest in periodic structures created on thin film and optical substrates has led to extensive studies of particular situations of propagation in periodic media. Araki and Itoh [42d], for instance, have investigated periodic ferrite slab waveguides whilst Lin et al [42e] have considered the case of periodic and quasi periodic gratings in optical dielectric waveguides. Tamir [42f] has also treated periodic optical waveguides. Finally Mickelson and Jaggard [42g] have presented a comprehensive treatment of electromagnetic wave propagation in almost periodic media, based upon the theory of the almost periodic Mathieu equation.

7.5 Electric Circuit Applications

Whenever an electric network has a component or indeed a complete branch of components varied in a periodic manner its characteristic equation in the time domain will be of the form of a periodic differential equation. Practical examples of such networks include parametric amplifiers, pulse modulation systems, sampled data filters and so-called N-path networks. In general the steady state forced response of such periodic systems will be described by (3.36) repeated here as

$$\boldsymbol{x}_{ss}(t) = \sum_{n=-\infty}^{\infty} \sum_{k=-\infty}^{\infty} v_{nk} \exp\{j(\omega_i + (n + k)\omega_p)t\}$$

where v_{nk} are weighting coefficients derived in Sect. 3.14, and ω_i and ω_p are the input and pump frequencies respectively. It is interesting and important to note in this expression that the steady state response contains components at the input frequency ($n = -k$ or $n, k = 0$) as well as 'switching harmonics' at $\omega_i + (n + k)\omega_p$. Whether these harmonics are of significant magnitude depends upon the relative sizes of the v_{nk}. These in turn depend upon network topology. In one particular application (such as a switched low pass filter) the network may be so chosen to cause the harmonics to be negligible whereas for a switching modulator a particular set of harmonics (i.e. sidebands) may be of paramount importance. The analysis of Sect. 3.14 has been developed for periodic systems in general. Analyses specific to circuit situations have been carried out by a number of authors; these essentially lead to expressions similar to that above in which the response frequencies are $\omega_i + (n + k)\omega_p$. Treatments of networks containing periodically operated switches, in particular, have been given by Fettweis [43], Liou [44] and more recently Strom and Signell [45]. Networks with sinusoidally varying components have been addressed by McDonald and Edmondson [46], Adams and Leon [47] and Kurth [48], with some more general aspects of periodic systems having been studied by Bardakjian and Sablatash [49] and Desoer [50]. The particular topic of stability in periodic networks has been considered by Orsic and Krajcinovic [51] for a second order circuit with an impulsively varying capacitance, by Sato [52] for a fourth order system, in which coupled parametric modes are of consideration, and by Sandberg [53]. Sandberg's work relates generally to any network which contains time-varying capacitors. Sandberg [54] has also looked at approximation methods for handling periodically time-varying non-linear networks. Despite many of these studies being of considerable interest in their own right, particularly from an applications viewpoint, detailed consideration will be given here only to two broad classes of periodic network. The first is that of parametric amplifiers and the second that of N-path networks. The former has been chosen because of its topicality and because of its classical appeal whereas the latter will be considered since the time-varying system equations arise not because of periodic parameters but rather as a result of multipliers or commutators imbedded in the networks.

In this section the essential properties of both parametric amplifiers and N-path networks will be deduced. Many practical aspects cannot be covered in the space available; for these, the treatment of parametric amplifiers by Howson and

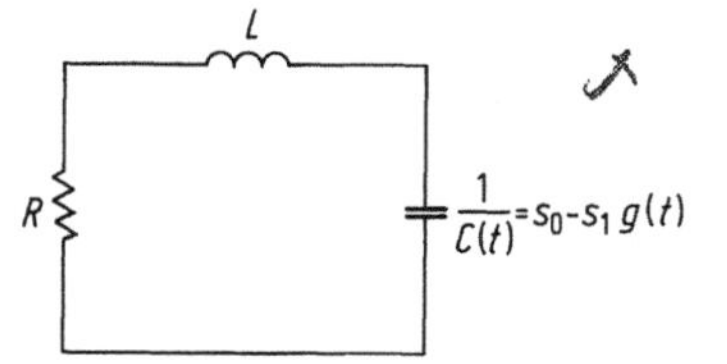

Fig. 7.5.1. Series *RLC* network with a periodically varying capacitance

Smith [55] will be seen to be of value, whilst the original paper by Franks and Sandberg [56], and the more recent text by Temes and Mitra [57] are recommended with respect to *N*-path circuits.

7.5.1 Degenerate Parametric Amplification

The simple series RLC network shown in Fig. 7.5.1 will be used to illustrate some basic facts concerning parametric amplification. The capacitor is considered to be varying periodically with time, according to

$$\frac{1}{C(t)} = s_0 - s_1 g(t) \tag{7.5.1}$$

where $g(t)$ is a periodic function of unit amplitude, and period θ seconds. Let $\omega_p = 2\pi/\theta$ be the fundamental radian frequency of the capacitor variation. This variation is caused, in practice, by an external source of energy. It is referred to as the *pump* source. In practice the pump source consists of a driven electrical network since the capacitor is usually the voltage dependent junction capacitance of a reverse biased semi-conductor diode called a *varactor*. Application of a pump voltage causes the capacitance to vary periodically in the manner described by (7.5.1).

The charge on the capacitor is described by the network equation

$$\frac{d^2q}{dt^2} + \frac{R}{L}\frac{dq}{dt} + \frac{1}{L}[s_0 - s_1 g(t)]q = 0. \tag{7.5.2}$$

It is important, for simplicity in the description, to use an equation in charge. Should a voltage equation be derived it will lead to additional first derivative terms arising from the effect of the time variable capacitance when the law $i(t) = \frac{d}{dt}q(t) = \frac{d}{dt}[C(t)\cdot v(t)]$ is applied.

For expediency assume that the series resistance in the network can be neglected, thereby allowing the first derivative term in (7.5.2) to be ignored. If R cannot be assumed negligible then (2.4) can be used to transform out the resulting first derivative.

Consider now the change of time variable

$$\omega_p t = 2\zeta. \tag{7.5.3}$$

Applying this, transforms (7.5.2), with $R = 0$, into

$$\frac{d^2q}{d\zeta^2} + \frac{4}{\omega_p^2 L}[s_0 - s_1 g(\zeta)]q = 0$$

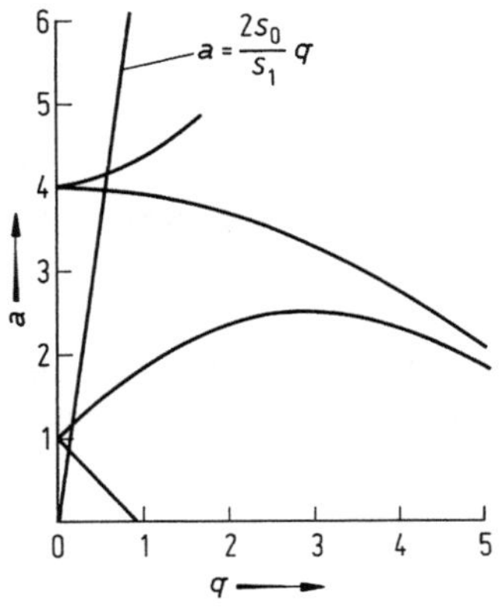

Fig. 7.5.2. Hypothetical stability diagram for the network of Fig. 7.5.1. showing the line $a = 2s_0/s_1\ q$ along which the circuit operates

where $g(\zeta)$ now has a period of π and a fundamental radian frequency of $2\ \text{rad} \cdot \text{s}^{-1}$. It is in the canonical form of a second order Hill equation (see (2.5)) with

$$a = \frac{4s_0}{\omega_p^2 L} = \left(\frac{2\omega_0}{\omega_p}\right)^2 \tag{7.5.4a}$$

$$q = \frac{2s_1}{\omega_p^2 L} = \left(\frac{2\omega_0}{\omega_p}\right)^2 \frac{s_1}{2s_0} \tag{7.5.4b}$$

where ω_0 is the natural frequency of the static, or unpumped, network. Note that a variation in pump frequency ω_p changes both a and q whilst keeping their ratio constant. As a result the natural behaviour of the simple pumped network, for a wide range of pump frequencies, can be ascertained by plotting a line of slope

$$\frac{a}{q} = 2s_0/s_1$$

on the corresponding stability diagram as shown in Fig. 7.5.2. From this it is seen readily that the charge in the network and thus the current, and the capacitor voltage, can be made to grow without bound if

$$a = \left(\frac{2\omega_0}{\omega_p}\right)^2 \simeq n^2$$

where n is an integer. Thus 'unbounded' amplification i.e. instability—can be invoked by pumping the capacitor at a rate

$$\omega_p = \frac{2\omega_0}{n}. \tag{7.5.5}$$

It should be noted that in the absence of any applied signal source it will be residual charge (noise) in the network that will be so amplified.

For parametric amplification of the type just described the pump frequency has to be defined exactly by (7.5.5) only if the corresponding value of q is small. It is evident from an inspection of Fig. 7.5.2 that larger q values (corresponding to larger capacitance variations) will invoke instability (and thus 'gain') for a

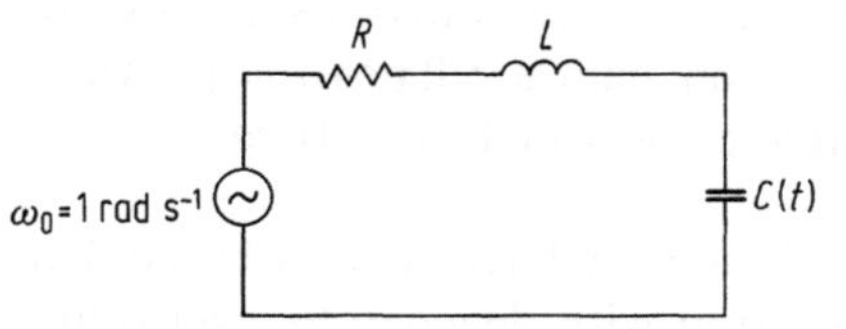

Fig. 7.5.3. Series *RLC* network with a periodically varying capacitance, forced at $\omega_0 = 1\ \text{rad} \cdot s^{-1}$.

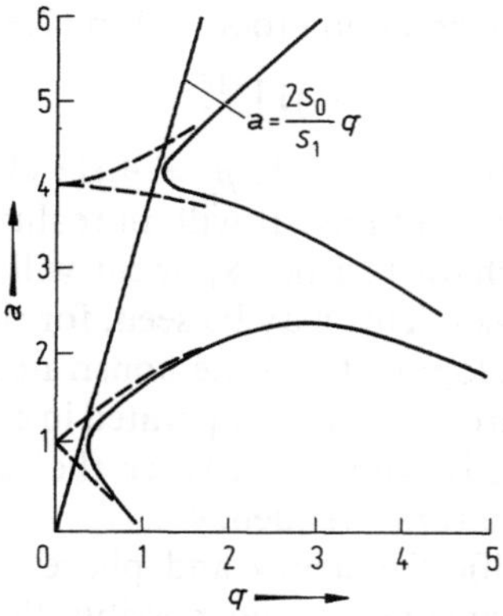

Fig. 7.5.4. Stability diagram for the lossy network of Fig. 7.5.3.

range of pump frequencies about those given in (7.5.5). Further it must be remarked that a specific phase relationship must exist between the pump and circuit signal (charge) waveforms. This is evident from the introductory discussions of Sect. 1.1.

In practice the circuit of Fig. 7.5.1 is of little value since it does not include a signal source. Moreover without a loss term the circuit would be unstable, and thus of no practical value, if pumped at $a = 1, 4, 9$ and so on. Fig. 7.5.3 shows the same circuit with an input generator at frequency ω_0. The resistance is now not considered zero but, for simplicity, is assumed sufficiently small that its value does not influence the value of $\omega_0 = \sqrt{s_0/L} = 1\ \text{rad} \cdot \text{s}^{-1}$.[3] Owing to the presence of the resistance the stability diagram is modified as depicted in Fig. 7.5.4. We assume that pumping is so arranged that the line describing the effect of a range of pump frequencies lies just on the stable side of all the stability boundaries. A rough analysis will now demonstrate that the circuit will provide large but stable outputs if the capacitor is pumped at twice the signal input frequency.

From (3.36) it can be seen that the response of the circuit pumped at ω_p and excited or forced at ω_0 will contain components at frequencies $\omega_0 + q\omega_p$ where q is an integer. The relative magnitudes of these components are given by the constants v_{nk} for which

$$v_{nk} \propto [-\mu + j(\omega_0 + k\omega_p)]^{-1}$$

as can be seen in the development leading up to (3.35); k is an integer that can take on all positive and negative values. For the special case of $k = -1$, $\omega_0 = 1$,

[3] Normalised $\omega_0 = 1$ and $\omega_p = 2$ are chosen to allow the properties of a stability diagram to be used.

$\omega_p = 2$ (corresponding to $n = 1$ in (7.5.5)) we have

$$v_{nk} \propto [-\mu - \mathrm{j}1]^{-1}.$$

In the vicinity of $a = 1$, $\mu = \pm\mathrm{j}1$; choosing the negative member shows that v_{nk} approaches infinity. It will, in reality, be bounded since the apparently purely imaginary characteristic exponent will have a negative real term resulting from network losses. This may be seen, for illustration, in Sect. 6.4.2.

This development has demonstrated in principle that, notwithstanding the fact that a circuit has to be operated in a stable region, gain is still possible provided the network is pumped at twice the input frequency, which in this case is also the system natural frequency.

The specific frequency and phase relationships that have to be observed in such an arrangement (and possibly the association with degenerate eigenvalues of the discrete transition matrix on stability boundaries) leads to the circuit being referred to as a *degenerate* parametric amplifier.

7.5.2 Degenerate Parametric Amplification in High Order Periodic Networks

The condition in (7.5.5) for a second order network is generalised in Sect. 4.4.3 to

$$\omega_p = \frac{2\omega_i}{n} \tag{7.5.6}$$

where ω_i is any one of the v natural frequencies of a system of order $2v$. For a fourth order system, for example, there will be two natural frequencies and thus two sets of pump frequencies that can be exploited for parametric gain. The properties of such a scheme have been partly explored by Cristaudo and Richards [58].

7.5.3 Nondegenerate Parametric Amplification

Fig. 7.5.5 shows the circuit of Fig. 7.5.1 redrawn to emphasise the practical point that an effective time-varying capacitance is created by pumping a non-linear capacitance (such as the voltage dependent junction capacitance of a varactor diode) by an externally applied sinusoidal voltage at the required frequency. It is necessary that the pump circuit contain a filter so that it will not been seen by the input generator which drives the left hand loop. In addition of course the pump circuit will only see the capacitance and not the remainder of the left hand input loop owing to its being tuned to ω_i, which can be quite different to ω_p. To ensure pump signals to not circulate in the input loop a well-defined narrow band filter is often used in the input.

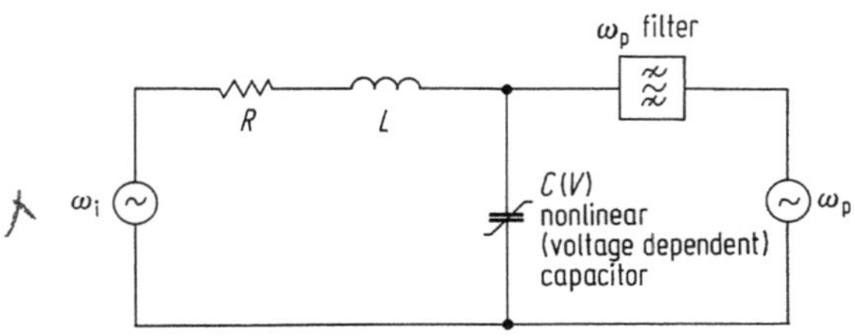

Fig. 7.5.5. Forced, series *RLC* network in which a periodically time-varying capacitance is created by driving a non-linear capacitance (such as the junction capacitance of a semiconductor diode) with a periodic potential

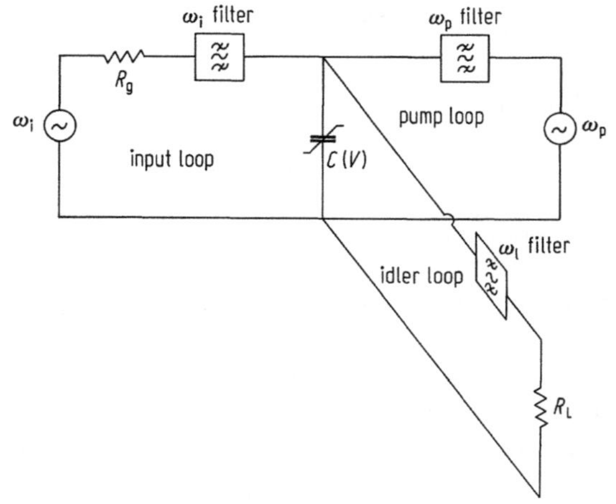

Fig. 7.5.6. Three frequency parametric circuit, consisting of signal, pump and idler loops

To allow the circuit to be considered as a time periodic, linear network two assumptions are made. First the signal on the capaciter at the input frequency is considered to be much lower in amplitude than that from the pump. This ensures that the capacitor's time variation is established by the pump alone. Notwithstanding this it is assumed also that the pump magnitude is small enough to ensure that the capacitance can be regarded as a time-varying linear element, and that non-linear effects are thus avoided. Consequently if the pump source is sinusoidal then so will be the capacitor variation. This will now be taken to be the case for the remainder of this section.

From (3.36) it is evident that the voltage across the capacitor is a complicated function, consisting of various amounts of components of all frequencies given by

$$\omega_i + q\omega_p,\ q = 0,\ \pm 1,\ \pm 2,\ \text{etc.} \tag{7.5.7}$$

Suppose we wished to extract a signal from the circuit at one of the frequencies available at the capacitor. To do this we would connect a load resistance across the capacitor, isolated by its own filter, as shown in Fig. 7.5.6. For reasons that will become clearer later this output loop is often referred to as the *idler circuit* or *loop*, tuned to a frequency ω_l. The circuit of Fig. 7.5.6 is the basic, so-called *three frequency parametric circuit.* We will proceed to show now that it can be used as an amplifier or as a frequency converter and amplifier, the distinction depending upon the specific choice of the idler frequency $\omega_l = \omega_i + q\omega_p$. Before proceeding it should be noted that a number of idler loops could be connected across the capacitor, each tuned to different values of q in (7.5.7).

7.5.4 Parametric Up Converters

The idler frequency for the circuit in Fig. 7.5.6 is given from (3.36) as

$$\omega_l = \omega_i + q\omega_p.$$

The 'conjugate' of this frequency is

$$\omega_l = -(\omega_i + q\omega_p)$$

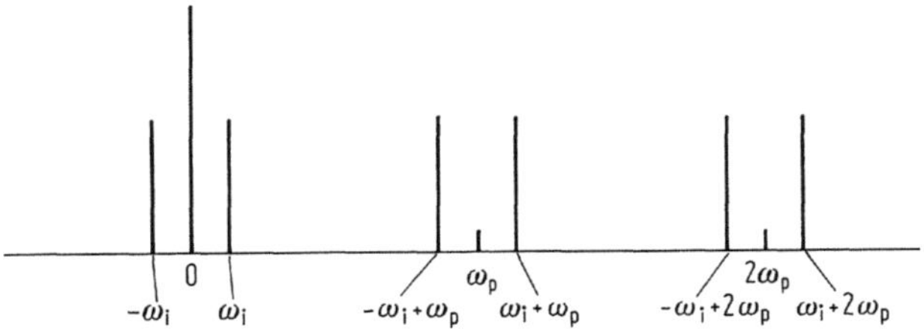

Fig. 7.5.7. Frequency spectrum for a parametric up-converter circuit

which would have appeared in (3.36) had the exponent in (3.30) been chosen negative. Consequently the frequency spectrum for the voltage across the capacitor appears as shown in Fig. 7.5.7 where it has been assumed that $\omega_p > \omega_i$. In this case the circuit is said to be an *up-converter*. Notice particularly the upper side band and lower side band designations on Fig. 7.5.7. These will be quite significant shortly.

It is expedient now to introduce an impedance matrix for the time-varying capacitance. For this let the capacitance be represented by

$$\begin{aligned} C(t) &= C_0(1 + 2\gamma \cos \omega_p t) \\ &= C_0(1 + \gamma e^{j\omega_p t} + \gamma e^{-j\omega_p t}). \end{aligned} \tag{7.5.8}$$

As a result of the filters in Fig. 7.5.6 the capacitor *current* only contains frequency components at the input and idler frequencies and can thus be represented

$$i(t) = I_i e^{j\omega_i t} + I_i^* e^{-j\omega_i t} + I_l e^{j\omega_l t} + I_l^* e^{-j\omega_l t}. \tag{7.5.9}$$

(The pump current has been taken care of implicitly in the capacitor expression).

Similarly the voltage across the capacitor of interest is of the form

$$v(t) = V_i e^{j\omega_i t} + V_i^* e^{-j\omega_i t} + V_l e^{j\omega_l t} + V_l^* e^{-j\omega_l t}. \tag{7.5.10}$$

In the above expressions the upper case letters represent phasor quantities. Equations (7.5.8), (7.5.9) and (7.5.10) will be used in the expression

$$i(t) = \frac{d}{dt}[C(t)v(t)]. \tag{7.5.11}$$

It is useful however to examine first the product $C(t)v(t)$,

$$\begin{aligned} C(t)v(t) = C_0[v(t) &+ \gamma V_i e^{j(\omega_p+\omega_i)t} + \gamma V_i e^{j(\omega_i-\omega_p)t} \\ &+ \gamma V_i^* e^{j(\omega_p-\omega_i)t} + \gamma V_i^* e^{j(-\omega_i-\omega_p)t} \\ &+ \gamma V_l e^{j(\omega_l+\omega_p)t} + \gamma V_l e^{j(\omega_l-\omega_p)t} \\ &+ \gamma V_l^* e^{j(\omega_p-\omega_l)t} + \gamma V_l^* e^{j(-\omega_p-\omega_l)t}]. \end{aligned} \tag{7.5.12}$$

At this stage attention has to be given to the choice of q in the idler frequency equations and thus specifically to the frequency handled by the idler circuit. Suppose for example we take the case of a *lower-side-band up converter*, for which the idler frequency is seen from Fig. 7.5.7 to be

$$\omega_l = \pm(\omega_i - \omega_p).$$

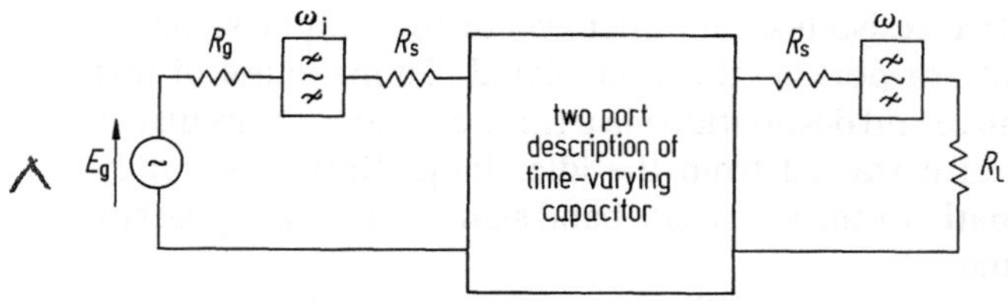

Fig. 7.5.8. Time-varying capacitance represented in two port matrix form imbedded in a source/load network. The left hand (input) loop is at frequency ω_i whilst the right hand (output) loop is at frequency ω_l. Note that the series loss resistance R_s of the time-varying capacitor has been referred to both frequencies

For such a circuit only terms with frequencies of $\pm\omega_i$, ω_p or $\pm(\omega_i - \omega_p)$ can exist in (7.5.12). Eliminating other terms and substituting into (7.5.11), along with (7.5.10) leads to the following pair of equations:

$$j\omega_i(C_0 V_i + \gamma C_0 V_l^*) = I_i$$

$$-j\omega_l(C_0 V_l^* + \gamma C_0 V_s) = I_l^*$$

where coefficients of $e^{j\omega_s t}$, $e^{-j\omega_l t}$ have been equated. These equations can be expressed as an admittance matrix for the time-varying capacitor that relates currents at the input and idler frequencies to voltages at those frequencies:

$$\begin{bmatrix} I_l^* \\ I_i \end{bmatrix} = \begin{bmatrix} -j\omega_l C_0 & -j\omega_l \gamma C_0 \\ j\omega_i \gamma C_0 & j\omega_i C_0 \end{bmatrix} \begin{bmatrix} V_l^* \\ V_i \end{bmatrix}.$$

This can be inverted to provide an impedance matrix. Assuming $\gamma^2 \ll 1$.

$$\begin{bmatrix} V_i \\ V_l^* \end{bmatrix} = \begin{bmatrix} \dfrac{1}{j\omega_i C_0} & \dfrac{\gamma}{j\omega_l C_0} \\ \dfrac{-\gamma}{j\omega_i C_0} & \dfrac{-1}{j\omega_l C_0} \end{bmatrix} \begin{bmatrix} I_i \\ I_l^* \end{bmatrix} = \begin{bmatrix} z_{11} & z_{12} \\ z_{21} & z_{22} \end{bmatrix} \begin{bmatrix} I_i \\ I_l^* \end{bmatrix}. \qquad (7.5.13)$$

With the time-varying capacitance described in this manner the circuit of Fig. 7.5.6 can be represented as shown in Fig. 7.5.8. From this representation it is evident that the input impedance of the complete circuit, seen at the input loop terminals, is

$$z_{\text{in}} = \frac{E_g}{I_i} = z_{11} + R_s + R_g - \frac{z_{12} z_{21}}{z_{22} + R_s + R_L}$$

where R_s is the series loss resistance of the capacitor. We can remove z_{11} and z_{22} from this expression if it is assumed they have been designed into the loop filters, having noted that they are the impedance of the static value of the capacitance at the input and idler frequencies. Thus, in practice

$$z_{\text{in}} = R_s + R_g - \frac{\gamma^2}{\omega_i \omega_l C_0^2 (R_s + R_L)}$$

which can be negative for small generator and capacitor loss resistance. The circuit therefore behaves as a negative resistance amplifier where the output signal is available at the same terminals, or port, as the input signal. To make use of this it is connected to the signal source and destination via a three or four port circulator [55]. Note that a signal is not extracted from the idler loop. Rather it simply sits in the circuit to provide a path for the lower side band signals. It is this apparent inactivity that leads to its name.

Consider now the case of an *upper side band up converter* for which

$$\omega_l = \pm(\omega_i + \omega_p).$$

With this choice of idler frequency a different set of terms are excluded from (7.5.12) leading to the impedance matrix description for the capacitor:

$$\begin{bmatrix} V_i \\ V_l \end{bmatrix} = \begin{bmatrix} \dfrac{-1}{j\omega_i C_0} & \dfrac{\gamma}{j\omega_l C_0} \\ \dfrac{\gamma}{j\omega_s C_0} & \dfrac{-1}{j\omega_l C_0} \end{bmatrix} \begin{bmatrix} I_s \\ I_l \end{bmatrix} = \begin{bmatrix} z_{11} & z_{12} \\ z_{21} & z_{22} \end{bmatrix} \begin{bmatrix} I_s \\ I_l \end{bmatrix}. \tag{7.5.14}$$

It can be shown that this arrangement will not present a negative input impedance to the signal source and thus cannot be used as a negative resistance amplifier. It does however provide gain from the input port to the idler circuit. In view of the differences in frequency it therefore functions as a frequency converter with gain. It is possible to show [55] that the maximum gain available is given by the ratio ω_l/ω_i.

7.5.5 *N*-Path Networks

N-path networks are periodically time-varying systems in which the variation arises not because of an explicit change in the value of a parameter but rather as a result of the incorporation of modulators or multipliers in the signal path. Switching sidebands similar to those in (3.36) arise, in this case owing to the multiplications which take place. Consequently they are related to amplitude-like modulation.

We will develop here the essential properties of *N*-path structures. The reader who is interested in more detail will find the development in Chap. 11 of Temes and Mitra [57] of value, from which the present treatment is adapted.

Consider the simple, single path network shown in Fig. 7.5.9 in which the input signal is multiplied by the periodic function $p(t)$ prior to its application to the time-invariant network of impulse response $h(t)$. The signal is extracted via a second multiplier in which it is multiplied by the function $q(t)$ which has the same period as $p(t)$. The signal applied to the time-invariant block is $x(t)p(t)$ and the output of that block is given by convolution as

$$\begin{aligned} z(t) &= h(t) * x(t)p(t) \\ &= \int_{-\infty}^{t} h(t-\tau)x(\tau)p(\tau)\mathrm{d}\tau. \end{aligned}$$

Thus the output of the complete system is

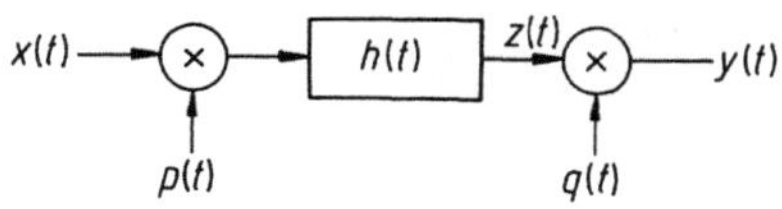

Fig. 7.5.9. Linear time invariant network with signals input and extracted via multipliers. This is the basic building block for N-path structures

$$y(t) = q(t) \cdot \int_{-\infty}^{t} h(t - \tau)x(\tau)p(\tau)\mathrm{d}r.$$
$$= \int_{-\infty}^{t} h(t - \tau)p(\tau)q(t)x(\tau)\mathrm{d}\tau. \tag{7.5.15}$$

We see by putting $x(\tau) = \delta(\tau)$ that the impulse response of the system of Fig. 7.5.9 is given by

$$W(t - \tau) = h(t - \tau)p(\tau)q(t). \tag{7.5.16}$$

Consider now the specific form of the output of the system, given from (7.5.15) for $x(t) = \mathrm{e}^{st}$ and $p(t)$ and $q(t)$ given by the complex Fourier series

$$p(t) = \sum_{m=-\infty}^{\infty} P_m \mathrm{e}^{\mathrm{j}2\pi mt/T}$$

and

$$q(t) = \sum_{l=-\infty}^{\infty} Q_l \mathrm{e}^{\mathrm{j}2\pi lt/T}.$$

Thus

$$y(t) = \sum_{m=-\infty}^{\infty} \sum_{l=-\infty}^{\infty} P_m Q_l \mathrm{e}^{\mathrm{j}2\pi lt/T} \int_{-\infty}^{t} h(t - \tau)\mathrm{e}^{\mathrm{j}2\pi m\tau/T}\mathrm{e}^{s\tau}\mathrm{d}\tau.$$

This can be rearranged as

$$y(t) = \sum_{m=-\infty}^{\infty} \sum_{l=-\infty}^{\infty} P_m Q_l \mathrm{e}^{\mathrm{j}2\pi(m+l)t/T}\mathrm{e}^{st}.$$
$$\int_{-\infty}^{t} h(t - \tau)\mathrm{e}^{-\mathrm{j}2\pi m(t-\tau)/T}\mathrm{e}^{-s(t-\tau)}\mathrm{d}\tau.$$

Introducing $\sigma = t - \tau$, we have $\mathrm{d}\tau = -\mathrm{d}\sigma$ and the integration limits become $-\infty$ to 0 so that

$$y(t) = \sum_{m=-\infty}^{\infty} \sum_{l=-\infty}^{\infty} P_m Q_l \mathrm{e}^{\mathrm{j}2\pi(m+l)t/T}\mathrm{e}^{st} \int_{0}^{\infty} h(\sigma)\mathrm{e}^{-(s+\mathrm{j}2\pi m/T)\sigma}\mathrm{d}\sigma.$$

The integral in the last expression is recognised as the single sided Laplace transform of the impulse response of the time-invariant block (i.e. its transfer function) but repeated periodically at the frequencies $2\pi m/T$. Thus the system output, expressed as

$$y(t) = \left[\sum_{m=-\infty}^{\infty} \sum_{l=-\infty}^{\infty} P_m Q_l \mathrm{e}^{\mathrm{j}2\pi(m+l)t/T} \cdot H(s + \mathrm{j}2\pi m/T)\right] \mathrm{e}^{st}$$

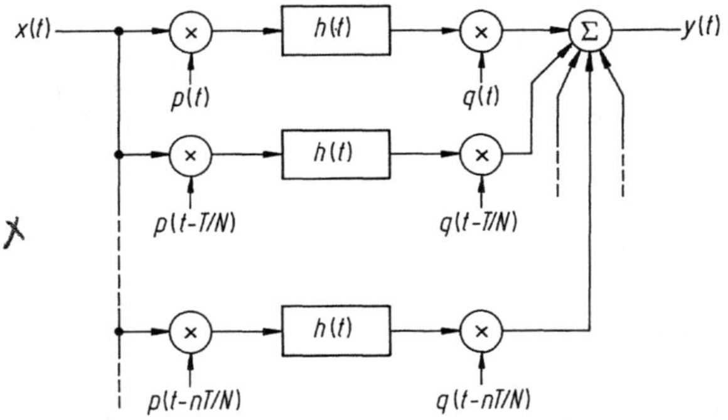

Fig. 7.5.10. The essential N-path network. The modulating functions $p(t-nT/N)$ and $q(t-nT/N)$ are distributed uniformly in phase over the modulating cycle of period T

is seen to consist of a set of components, each resulting from signal transmission through a time-invariant block, but translated to different harmonically related frequencies. If the input frequency is sinusoidal with $s = j\omega_i$ and the frequency of the modulating functions $2\pi/T$ is called ω_p then the output of the network of Fig. 7.5.9 will contain components at the frequencies $\omega_i + (m + l)\omega_p$ which are the same components that would be obtained from a network with a periodically varying parameter, as revealed by (3.36). The relative magnitudes of these components are determined by the complex Fourier coefficients P_m and Q_l, and the magnitude of $H(s)$ at $s = j\omega = (\omega_i + m\omega_p)$. It can be shown now that these sidebands can be removed by combining a number of the networks of Fig. 7.5.9 in parallel.

The network of Fig. 7.5.9 is the building block for the N-path structure, shown now in Fig. 7.5.10. The time-invariant blocks are all identical but the input and output modulating functions have different phases for each path. Indeed their phases are distributed uniformly as shown in the diagram. Consequently, for the nth path in the structure the complex Fourier series for the modulating functions are now given by

$$p(t - nT/N) = \sum_{m=-\infty}^{\infty} P_m e^{j2\pi mt/T} e^{-j2\pi mn/N}$$

and

$$q(t - nT/N) = \sum_{l=-\infty}^{\infty} Q_l e^{j2\pi lt/T} e^{-j2\pi ln/N}.$$

From (7.5.16) the impulse response of the nth path is therefore

$$W_n(t - \tau) = h(t - \tau) \sum_{m=-\infty}^{\infty} \sum_{l=-\infty}^{\infty} P_m Q_l e^{-j2\pi(m+l)n/N} \cdot e^{j2\pi m\tau/T} e^{j2\pi lt/T}.$$

The impulse response for the complete structure is given by

$$\begin{aligned} W(t - \tau) &= \sum_{n=0}^{N-1} W_n(t - \tau) \\ &= h(t - \tau) \sum_{m=-\infty}^{\infty} \sum_{l=-\infty}^{\infty} P_m Q_l \cdot e^{j2\pi(m\tau + lt)/T} \sum_{n=0}^{N-1} e^{-j2\pi(m+l)n/N}. \end{aligned}$$

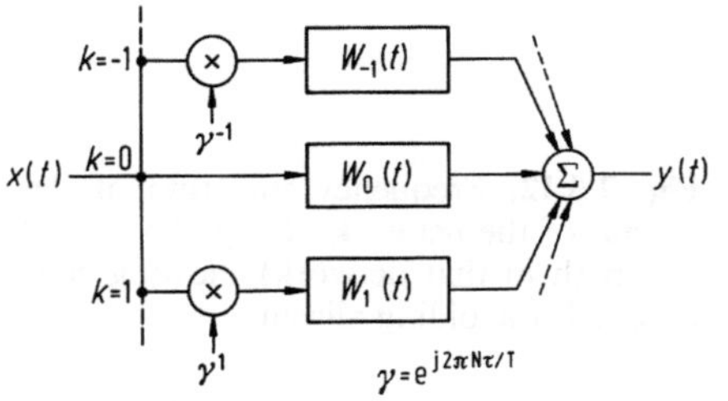

Fig. 7.5.11. Signal equivalent N-path structure

It is shown readily that

$$\sum_{n=0}^{N-1} e^{-j2\pi(m+l)n/N} = N \quad \text{for } m + l = kN \text{ where } k \text{ is an integer}$$
$$= 0 \quad \text{otherwise.}$$

Consequently, the impulse response can be recast as

$$\begin{aligned} W(t-\tau) &= Nh(t-\tau) \sum_{k=-\infty}^{\infty} \sum_{l=-\infty}^{\infty} P_{kN-l} Q_l \cdot e^{j2\pi kN\tau/T} \cdot e^{j2\pi l(t-\tau)/T} \\ &= \sum_{k=-\infty}^{\infty} e^{j2\pi kN\tau/T} W_k(t-\tau) \end{aligned} \tag{7.5.17}$$

where

$$W_k(t-\tau) = N \sum_{l=-\infty}^{\infty} P_{kN-l} Q_l h(t-\tau) e^{j2\pi l(t-\tau)/T}. \tag{7.5.18}$$

Equation (7.5.17) can be interpreted in terms of an abstract, signal N-path structure that is equivalent to the original, physically parallel structure as shown in Fig. 7.5.11. In this, the impulse responses of the time-invariant blocks in each path are the $W_k(t - \tau)$ of (7.5.18). Each of these is fed via an input modulator as shown, where the index k establishes the frequency onto which the input signal $x(t)$ is modulated. For the particular case of $k = 0$ the input is not frequency translated and indeed appears at its original frequency at the output, modified in amplitude and phase by the appropriate transform function ((7.5.18) with $k = 0$). Transmission of the input signal via the $k \neq 0$ paths in Fig. 7.5.11 leads to the production of the modulation sidebands met earlier with respect to Fig. 7.5.10, although here there are $(N - 1)$ times as many. These sidebands will not appear in the output signal if either of two conditions apply.

(i) If $P_{kN-l}Q_l = 0$ for all $k \neq 0$ then the only signal transmission path in Fig. 7.5.11 is that for $k = 0$. This requires

$$P_{-l} = Q_l = 0 \quad \text{for } |l| > N/2$$

and implies that the modulating functions must be appropriately band limited.

(ii) If the input signal contains no components with frequencies greater than $f_c = N/2T$ and the output of the N-path is extracted via a low pass filter of cut-off frequency $N/2T$ then only the signal through $k = 0$ will appear at the output

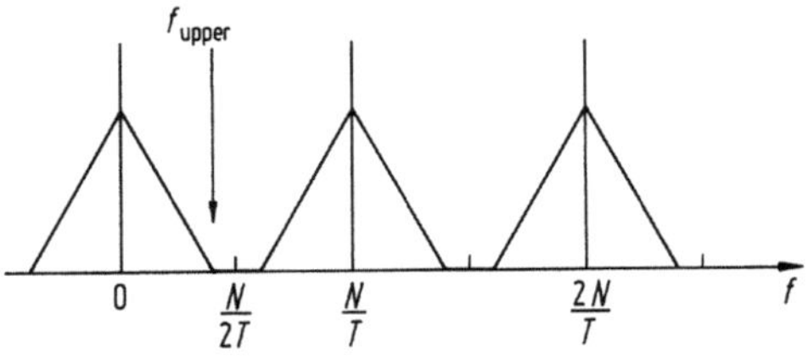

Fig. 7.5.12. Frequency spectrum of the output of the network of Fig. 7.5.11. (or one path in that network) showing conditions for avoiding aliasing

of the filter. To see this note that the frequency spectrum of the output $y(t)$ in Fig. 7.5.11 is as shown in Fig. 7.5.12 where the input signal is band limited to f_{upper}. Extracting the output via the filter described excludes all the images of $x(t)$. If $x(t)$ is not suitably band limited aliasing occurs and the output becomes distorted by the sideband images interfering with the $k = 0$ signal. To ensure this does not occur the N-path structure can be preceded by an $N/2T$ (antialiasing) low pass filter as well as being followed by one.

If either of the above conditions hold then the impulse response of the N-path network reduces simply to

$$W(t - \tau) = N \sum_{l=-\infty}^{\infty} P_{-l} Q_l h(t - \tau) e^{j2\pi l(t-\tau)/T} \tag{7.5.19}$$

which is a periodic or comb like transfer function repeated around the frequencies $f = l/T$ as depicted in Fig. 7.5.13. This does not mean that outputs at each of those frequencies are produced for a single input frequency (as is the case for the $k \neq 0$ paths as discussed above) but rather that $W(t - \tau)$ is frequency selective in a periodic fashion. This allows the N-path structure, via a manipulation of the P_l and Q_l, to be used as a highly selective band pass filter, a comb filter and so on, applications that are discussed at length in Temes and Mitra [57].

Sun and Frisch [59] have presented an analysis for a more general N-path of the type in Fig. 7.5.10 in which the time-invariant impulse responses $h(t)$ are all different but in which the modulators are ring-commutators which share the input sequentially among the paths.

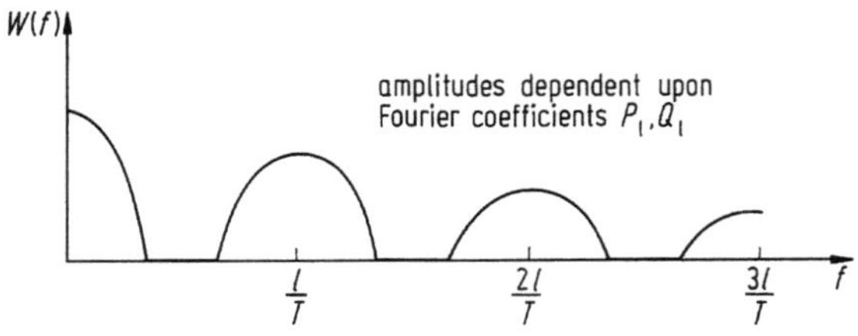

Fig. 7.5.13. Transfer function of the $k = 0$ path in the structure of Fig. 7.5.11. showing its comb-like nature. The relative amplitudes and phases of the individual pass bands are determined by the P_l, Q_l and exponential factors

References for Chapter 7

1. McLachlan, N. W.: Theory and application of Mathieu functions. Clarendon: Oxford U.P. 1947. Reprinted by Dover, New York 1964
2. Courant, E. D.; Livingston M. S.; Snyder, H. S.: The strong-focussing synchrotron—a new high energy accelerator. Phys. Rev. 88 (1952) 1190–1196

3. Paul, W.; Steinwedel, H.: Ein neues Massenspektrometer ohne Magnetfeld. Z. Naturforsch. 8a (1953) 448–450
4. Blauth, E. W.: Dynamic mass spectrometry. Amsterdam: Elsevier 1966
5. Kiser, R. W.: Introduction to mass spectrometry and its applications. Englewood Cliffs: Prentice Hall 1965
6. Dawson, P. H.: Quadrupole mass spectrometry. Amsterdam: Elsevier 1976
7. Denison, D. R.: Operating parameters of a quadrupole in a grounded cylindrical housing. J. Vac. Sci. Technol. 8 (1971) 266–268
8. Richards, J. A.; Huey, R. M.; Hiller, J.: A new operating mode for the quadrupole mass filter. Int. J. Mass Spectrom. Ion Phys. 12 (1973) 317–339
9. Richards, J. A.; Huey, R. M.; Hiller, J.: Conditions for selectivity in a quadrupole mass filter. Int. J. Mass Spectrom. Ion Phys. 15 (1974) 379–390
10. Richards, J. A.: Conditions for selectivity in a quadrupole mass filter: Addendum. Int. J. Mass Spectrom. Ion Phys. 17 (1975) 213–214
11. von Zahn, U.: Monopole spectrometer—a new electric field mass spectrometer. Rev. Sci. Instrum. 34 (1963) 1–4
12. Hudson, J. B.; Watters, R. L.: The monopole—a new instrument for measuring partial pressures. IEEE Trans. Instrum. Meas. IM-15 (1966) 94–98
13. Lawson, G.; Bonner, R. F.; Todd, J. F. J.: The quadrupole ion store (quistor) as a novel source for a mass spectrometer. J. Phys. E Sci. Instrum. 6 (1973) 357–362
14. Lever, R. F.: Computation of ion trajectories in the monopole mass spectrometer by numerical integration of Mathieu's equation. IBM J. Res. Dev. 10 (1966) 26–40
15. Dawson, P. H.; Whetten, N. R.: The monopole mass spectrometer. Rev. Sci. Instrum. 39 (1968) 1417–1422
16. Dawson, P. H.; Whetten, N. R.: Some properties of the monopole mass spectrometer. J. Vac. Sci. Technol. 6 (1969) 97–99
17. Dawson, P. H.; Whetten, N. R.: Ion storage in three-dimensional rotationally symmetric quadrupole fields. I. Theoretical treatment. J. Vac. Sci. Technol. 5 (1968) 1–10
18. Richards, J. A.: An improved quadrupole mass spectrometer. PhD Thesis, University of New South Wales, Kensington, Australia, 1972
19. Pipes, L. A.: Matrix solution of equations of the Mathieu-Hill type. J. Appl. Phys. 24 (1953) 902–910
20. Dawson, P. H.: A detailed study of the quadrupole mass filter. Int. J. Mass Spectrom. Ion Phys. 14 (1974) 313–337
21. Dawson, P. H.; Lambert, C.: A detailed study of the quadrupole ion trap. Int. J. Mass Spectrom. Ion Phys. 16 (1975) 269–280
22. Richards, J. A.; Huey, R. M.; Hiller, J.: The truncated Hill determinant as a tool in quadrupole mass filter studies. Int. J. Mass Spectrom. Ion Phys. 13 (1974) 443–451
23. Richards, J. A.: On the choice of steps in the piece wise-constant Hill equation model of a quadrupole mass filter. Int. J. Mass Spectrom. Ion Phys. 18 (1975) 11–19
24. Richards, J. A.; McLellan, R. N.: Fast computer simulation of a quadrupole mass filter driven by a sinusoidal RF waveform. Int. J. Mass Spectrom. Ion Phys. 17 (1975) 17–22
24a. Bonner, R. F.; Hamilton, G. F.; March, R. E.: Calculation of the phase space parameters for the study of quadrupole devices. Int. J. Mass Spectrom. Ion Phys. 30 (1979) 365–371
24b. Baril, M.: Etude des properties fondomentales de l'equation de Hill pour le dessin de filtre quadrupolaire. Int. J. Mass Spectrom. Ion Phys. 35 (1980) 179–200
24c. Dawson, P.: Ion optical data for the quadrupole mass filter. Int. J. Mass Spectrom. Ion Phys. 36 (1980) 353–364
24d. Todd, J. F. J.; Freer, D. A.; Waldren, R. M.: The quadrupole ion store (Quistor). Part XI: The model of ion motion in a pseudo potential well: An appraisal in terms of phase-space dynamics. Int. J. Mass Spectrom. Ion Phys. 36 (1980) 185–203.
24e. Todd, J. F. J.; Freer, D. A.; Waldren, R. M.: The quadrupole ion store (Quistor). Part XII: The trapping of ions injected from an external source: A description in terms of phase space dynamics. Int. J. Mass Spectrom. Ion Phys. 36 (1980) 371–386
24f. Chun-Sing, O; Schwessler, H. A.: Confinement of pulse-injected external ions in a radiofrequency quadrupole ion trap. Int. J. Mass Spectrom. Ion Phys. 40 (1981) 53–66

25. Dawson, P. H.; Whetten, N. R.: Quadrupole mass filter: Circular rods and peak shapes. J. Vac. Sci. Technol. 7 (1970) 440–441
26. Dawson, P. H.: Use of fringing fields in the monopole. J. Vac. Sci. Technol. 8 (1971) 263–265
27. Dawson, P. H.; Whetten, R. N.: Nonlinear resonances in quadrupole mass spectrometers due to imperfect fields. I. The quadrupole ion trap. Int. J. Mass Spectrom. Ion Phys. 2 (1969) 45–49
28. Dawson, P. H.; Whetten, N. R.: Nonlinear resonances in quadrupole mass spectrometers due to imperfect fields. II. The quadrupole filter and monopole mass spectrometer. Int. J. Mass Spectrom. Ion Phys. 3 (1969) 1–12
29. Winters, J. L.; Musumeci, R. A.; Richards, J. A.: Optimum analysing electrode radius for a fourfold monopole mass spectrometer. Int. J. Mass Spectrom. Ion Phys. 31 (1979) 31–36
30. Timoshenko, S. P.; Gere, J. M.: Theory of elastic stability, 2nd ed. New York: McGraw-Hill 1961
31. Walker, G. R.: An investigation of the dynamic stability of a boom crane. Proc. Inst. Eng. (Aust) Conf. on metal structures research and its applications. Sydney, Nov. 1972
32. Karbowiak, A. E.: Trunk waveguide communication. London: Chapman & Hall 1965
33. Proc. IEEE 68, Oct. 1980. Special issue on optical fibre communications.
34. Kraus, J. D.; Carver, K. R.: Electromagnetics, 2nd ed. New York: McGraw-Hill 1973
34a. Rengarajan, S.; Lewis, J. E.: Dielectric loaded elliptical waveguides. IEEE Trans. Microwave Theory Tech. MTT-28 (1980) 1085–1089
34b. Rengarajan, S.; Lewis, J. E.: The elliptical surface wave transmission line. IEEE Trans. Microwave Theory Tech. MTT-28 (1980) 1089–1095
34c. Yamashita, E.; Atsuki, K.; Nishino, Y.: Composite dielectric waveguides with two elliptic-cylinder boundaries. IEEE Trans. Microwave Theory Tech. MTT-29 (1981) 987–990
35. Kretzschmar, J. G.: Wave propagation in hollow conducting elliptical waveguides. IEEE Trans. Microwave Theory Tech. MTT-18 (1970) 547–554
36. Brillouin, L.: Wave propagation in periodic structures. New York: McGraw-Hill 1946. Reprinted by Dover (2nd ed.), New York 1953
37. Elachi, C.: Wave in active and passive periodic structures: A review. Proc. IEEE 64 (1976) 1666–1698
37a. Chuang, S-L.; Kong, J. A.: Scattering of waves from periodic surfaces. Proc. IEEE 69 (1981) 1132–1144
38. Tamir, T.; Wang, H. C.; Oliner, A. A.: Wave propagation in sinusoidally stratified dielectric media. IEEE Trans. Microwave Theory Tech. MTT-12 (1964) 323–335
39. Yeh, H.; Casey, K. F.; Kaprielian, Z. A.: Transverse magnetic wave propagation in sinusoidally stratified dielectric media. IEEE Trans. Microwave Theory Tech. MTT-13 (1965) 297–302
40. de Kronig, R. L.; Penney, W. G.: Quantum mechanics of electrons in crystal lattices. Proc. R. Soc. (Ser. A) 130 (1931) 499–513
41. Ferry, D. K.; Fannin, D. R.: Physical electronics. Reading: Addison-Wesley 1971
42. Weinreich, G.: Solids: Elementary theory for advanced students. New York: Wiley 1965
42a. Seshadri, S. R.: Asymptotic theory of mode coupling in a space-time periodic medium—Part I: Stable interactions Proc. IEEE 65 (1977) 996–1004
42b. Seshadri, S. R.: Asymptotic theory of mode coupling in a space-time periodic medium—Part II: Unstable interactions Proc. IEEE 65 (1977) 1459–1469
42c. Seshadri, S. R.: Reflection and transmission coefficients of a periodic dielectric slab. Proc. IEEE 66 (1978) 699–700
42d. Araki, K; Itoh, T.: Analysis of periodic ferrite slab waveguides by means of improved perturbation method. IEEE Trans. Microwave Theory Tech. MTT-29 (1981) 911–916
42e. Lin, Z–Q.; Zhou, S.–T.; Chang, W. S. C.; Forouhar, S.; Delavaux, J. M.: A generalised two-dimensional coupled-mode analysis of curved and chirped periodic structures in open dielectric waveguides. IEEE Trans. Microwave Theory Tech. MTT-29 (1981) 881–891

42f. Tamir, T.: Microwave modelling of optical periodic waveguides. IEEE Trans. Microwave Theory Tech. MTT-29 (1981) 979–983
42g. Mickelson, A. R.; Jaggard, D. L.: Electromagnetic wave propagation in almost periodic media. IEEE Trans. Antennas Propag. AP-27 (1979) 34–40
43. Fettweis, A.: Steady-state analysis of circuits containing a periodically-operated switch. IRE Trans. Circuit Theory CT-6 (1959) 252–260
44. Liou, M.: Exact analysis of linear circuits containing periodically-operated switches, with applications. IEEE Trans. Circuit Theory CT-19 (1972) 146–154
45. Strom, T.; Signell, S.: Analysis of periodically switched linear circuits. IEEE Trans. Circuits Syst. CAS-24 (1977) 531–541
46. McDonald, J. R.; Edmondson, D. E.: Exact solution of a time-varying capacitance problem. Proc. IRE (1961) 453–465
47. Adams, J. V.; Leon, B. J.: Steady state analysis of linear networks containing a single sinusoidally varying capacitor. IEEE Trans. Circuit Theory CT-14 (1967) 313–319
48. Kurth, C. F.: Steady state analysis of sinusoidal time-variant networks applied to equivalent circuits for transmission networks. IEEE Trans. Circuits Syst. CAS-24 (1977) 610–624
49. Bardakjian, B. L.; Sablatash, M.: Spectral analysis of periodically time-varying linear networks. IEEE Trans. Circuit Theory CT-19 (1972) 297–299
50. Desoer, C. A.: Steady state transmission through a network containing a single time-varying element. IRE Trans. Circuit Theory CT-6 (1959) 249–252
51. Orsic, M.; Kajcinovic, D.: Resonant circuit with impulsively varying capacitance. IEE Trans. Circuit Theory CT-19 (1972) 533–536
52. Sato, C.: Stability conditions for resonant circuits with time-variable parameters. IRE Trans. Circuit Theory CT-9 (1962) 340–349
53. Sandberg, I. W.: A stability criterion for linear networks containing time-varying capacitors. IEEE Trans. Circuit Theory CT-12 (1965) 2–11
54. Sandberg, I. W.: On truncation techniques in the approximate analysis of periodically time-varying nonlinear networks. IEEE Trans. Circuit Theory CT-11 (1964) 195–201
55. Howson, D. P.; Smith R. B.: Parametric amplifiers. Maindenhead: McGraw-Hill 1970
56. Franks, L. E.; Sandberg, I. W.: An alternative approach to the realization of network tranfer functions: The *N*-path filter. Bell Syst. Tech. J. 39 (1960) 1321–1350
57. Temes, G. C.; Mitra, S. K. (Eds.): Modern filter theory and design. New York: Wiley 1973
58. Cristaudo, P. G.; Richards, J. A.: Conditions for a two channel degenerate parametric amplifier. 16th Convention of the Institution of Radio and Electronics Engineers, Australia, Melbourne, August 1977
59. Sun, Y.; Frish, I. T.: A general theory of commutated networks. IEEE Trans. Circuit Theory CT-16 (1969) 502–508

Problems

7.1 When ions are injected into a quadrupole mass filter they travel from the field free region of the ion source, through a fringing region in the vicinity of the end of the quadrupole electrode structure and then into the fully developed quadrupole field. Their passage through the fringing region can be examined on the stability diagram of Fig. 7.1.6 by observing that a and q in (7.1.10) increase from zero to the particular values for a given ion in accord with the increases experienced in U and V. Thus as an ion traverses the fringing region its 'operating point' moves along the mass scan line of Fig. 7.1.6, from the origin to its final value. Consequently it will have initially an unstable trajectory. Many ions are lost in this manner giving the mass filter a lower transmission than would otherwise be the case. Brubaker [Brubaker W. M.: An improved quadrupole mass analyser. Adv. Mass Spectrom. (1968)] has suggested that this problem may be overcome by preceding the main mass filter by an auxiliary quadrupole electrode set to which is applied only the ac components of the energising potentials. Explain why this is beneficial.

7.2 Using (7.1.12) show that the equations of motion for ions in the quadrupole ion trap of Fig. 7.1.11 are given by the single Mathieu equation

$$\ddot{x} + (a - 2q\cos 2t)x = 0$$

with

$$a = a_r = 8eU\eta_r$$

$$q = q_r = -4eV\eta_r, \quad \eta_r = (mr_0^2\omega^2)^{-1}$$

for radial motion, and

$$a = a_z = -6eU\eta_z$$

$$q = q_z = 8eV\eta_z, \quad \eta_z = (mz_0^2\omega^2)^{-1}$$

for axial motion. As a result sketch a composite stability diagram for the device.

7.3 Can you foresee dynamic buckling, as described in Sect. 7.2 to be a problem with the side rods of steam locomotives?

7.4 If more than one mode can propagate simultaneously in a waveguide distortion of the information being transmitted frequently occurs since the modes travel with different velocities. Consequently a waveguide is often operated in the frequency range between the cut-off frequency of its dominant mode and that of the next highest mode. The bandwidth available for use is that frequency range. Sketch a graph of the (normalised) bandwidth of an elliptical guide over a representative range of eccentricities.

7.5 The cut-off frequency of a waveguide containing a dielectric that varies periodically in the direction of propagation can be determined from the intersection of the line

$$a = -\frac{\varepsilon_1 q}{2\varepsilon(0)} - \left(\frac{h\theta}{\pi}\right)^2 \qquad \text{(see (7.4.14b, c))}$$

with the a_0 stability boundary of the appropriate stability diagram. This boundary can be approximated by

$$a = -2q^2\Psi_1\Psi_1^*$$

where Ψ_1 is the complex coefficient of the fundamental in the Fourier series that describes the periodic dielectric variation. (Assume a sinusoidal dielectric variation, for convenience). Determine an expression for the cut-off frequency of the guide and ascertain that this is no higher than that for the guide filled completely with a dielectric of value $\varepsilon(0)$.

7.6 A stack of very thin glass plates separated by very fine airgaps will preferentially transmit light of different colours. Explain this effect. This explanation was used by Lord Rayleigh to describe the coloured appearance of glass which had become laminated with age.

7.7 A hypothetical crystal lattice presents to electrons a periodic potential as shown in Prob. 7.13. If the effective mass of the electron is 9.1×10^{-31} kg, $h = 1.06 \times 10^{-35}$ Js and the lattice constant is 2.5×10^{-10} m, determine the size of the first forbidden zone relative to the lattice potential.

7.8 A simple LC series circuit consists of a 10 μH inductor and a junction diode capacitance of nominal value 5 pF. Determine the two highest frequencies at which instability can be induced in the circuit by pumping the capacitance. If the inductor has a series loss resistance of 10 Ω use a 3×3 Hill determinant to approximate the degree of pumping (q) required to induce instability at the highest suitable pump frequency if the pump waveform is of the form $(\cos\omega_p t)^{-1}$. Use an equation in charge for your analysis.

7.9 A filter of centre frequency 100 Hz and bandwidth 5 Hz is required for a particular application. Describe how such a filter can be synthesised using an N-path configuration.

7.10 Consider an array of electric quadrupole lenses of the type shown in Fig. 7.1.1, arranged with successive 90° rotations so that a charged particle injected axially through the array will experience alternating electric fields in both the x and y directions with only dc excitation applied to the lenses. The potential distribution experienced by a particle therefore will be of the form

$$V = \frac{U_0}{r_0^2}(y^2 - x^2)F(z)$$

where U_0 is the applied potential, r_0 is half the inter-electrode spacing (so-called "field radius") and $F(z)$ is a unit square wave function of spatial period L, L being the spacing between the centres of the lenses. If a particle of mass m and charge e is injected into the structure with a constant velocity v show that its motion is described by Meissner's equation. If the initial energy of the ion is E, show that the structure can be used as an energy filter.
(See Taubert, R.: Ein elektrostatisches Energiefilter. Z. Naturforsch. 12a (1957) 169.

7.11 A second order system with the (frequency domain) transfer function

$$A(s) = (s^2 + 2\zeta s + \omega_0^2)^{-1}$$

has negative feedback applied around it. Imagine the feedback path contains a periodically operated switch which is operated with period T and is closed for τ seconds each cycle. When the switch is closed the feedback factor seen by the system is β; otherwise it is zero.

Show that the system is described by a Hill equation with a rectangular coefficient and thereby demonstrate that it is prone to instability if the switch closure time τ and the feedback fraction β are not carefully chosen. Transform the equation to the form of a standard Meissner equation and identify the regions of the stability diagrams of Fig. 4.2 and 4.3 of significance.

Consider the implication of changing the feedback to positive feedback.
(See Hiller, J.; Keenan, R. K.: Stability of finite-width, sampled-data systems. Int. J. Control 8 (1968) 1–22 and, Tait, K. E.; Sutton, G. R.: The determination of stability boundaries and control energy for a finite pulse width sampled data system. Int. J. Control 3 (1966) 229–254.

7.12 The operation of the quadrupole mass spectrometers described in Sect. 7.1 becomes severely impaired if the applied periodic potential contains odd multiples of the first sub-harmonic, and yet is not impaired by harmonics of the potential. Why should this be so? You may find Floquet theory of value in addressing this issue.
(See von Busch, F.; Paul, W.: Über nichtlineare Resonanzen in elektrischen Massenfilter als Folge von Feldfehlern. Z. Phys. 164 (1961) 588–594.

7.13 The derivation of Schrödinger's equation in Sect. 7.4.5 describing electrons in crystal lattices has assumed that the periodic lattice potential is alternately and equally positive and negative. In reality this is not the case; instead a rectangular periodic potential consistent with a crystal lattice is illustrated below. It is unipolar and has a dc or constant term that depends upon its amplitude. How would the use of this potential in (7.4.15) modify the essential results developed in Sect. 7.4.5.?

Appendix

Bessel Function Generation by Chebyshev Polynomial Methods

One possible approach to generating the Bessel functions required in Sect. 3.6 to 3.10 is to evaluate the series expansions which serve as their definitions. From the point of view of numerical evaluation this is undesirable owing to the required fractional exponentiation for functions of fractional order. Furthermore it is difficult, with this method, to obtain an estimate of the error made by truncating the series at a particular term. A more efficient technique is to employ a Chebyshev polynomial expansion of the desired function in the form

$$f(x) = \sum_{r=0}^{\infty}{}' a_r T_r(x) \tag{A1}$$[1]

where the Chebyshev polynomial, $T_r(x)$, defined by

$$T_r(x) = \cos(r \cos^{-1} x), \quad -1 \leq x \leq 1$$

is subject to the recurrence relation

$$T_{r+1}(x) - 2xT_r(x) + T_{r-1}(x) = 0. \tag{A2}$$

An approximation $g(x)$, to a function $f(x)$, can be produced by truncating the expansion such that

$$g(x) = \sum_{r=0}^{n}{}' a_r T_r(x) \tag{A3}$$

where the error incurred by truncation has an upper bound given simply by

$$\varepsilon_n(x) = \sum_{r=n+1}^{\infty} |a_r|.$$

Owing to the rapid convergence of the series, a desired accuracy (and thus the value of n) can be chosen simply by inspection of the available range of expansion coefficients. As a result of this feature, and the fact that tables of Chebyshev polynomial expansion coefficients are readily available, this approach is an attractive practical one. Clenshaw and Picken [1] have produced tables of Chebyshev expansion coefficient for Bessel functions of fractional order. The coefficients are given to the 20th decimal place.

---

[1] The prime on the summation indicates that the leading term in the series is halved.

The more usual method of carrying out a Chebyshev expansion is to make use of the recurrence relation in Eq. (A2). Since $T_0(x) = 1$ and $T_1(x) = x$ the higher polynomials are generated quite readily and may be incorporated into an algorithm for machine evaluation. A disadvantage with this approach lies in the number of arithmetic operations necessary to evaluate a given number of terms in the series. If computer time rationing is not a major consideration, the accompanying advantages of flexibility and accuracy offset slow computation. The technique is flexible in that a change in order of the approximation simply requires the addition of more terms to the expansion whilst accuracy is guaranteed by the properties of Chebyshev expansions.

The alternative to direct evaluation of the Chebyshev polynomials involves the rearrangement of a given Chebyshev series into a power series. This power series may then be evaluated by the well known technique of nested multiplication, leading to rapid computation. A requirement of this method is that the power series expansion coefficients must be generated from the Chebyshev coefficients. Whilst this in itself is particularly simple, it does represent a loss of flexibility since a change in the order of the approximation to a function requires a new set of power series coefficients. Another limitation is the possible ill-determination of the power series coefficients, thus leading to some uncertainty as to the accuracy of the technique. This is discussed in Clenshaw [2].

The accuracy of the Bessel function values generated by Chebyshev methods can be checked against the 10 decimal place tables produced by the National Bureau of Standards [3, 4].

References for Appendix

1. Clenshaw, C. W.; Picken, S. M.: Chebyshev series for Bessel functions of fractional order. N.P.L. Mathematical tables, Vol. 8. London: H. M. Stat. Office 1966
2. Clenshaw, C. W.: Chebyshev series for mathematical functions. N.P.L. Mathematical tables, Vol. 5 London: H. M. Stat. Office 1962
3. Nat. Bur. Stand. Computation Laboratory: Tables of Bessel functions of fractional order, Vol. 1. New York: Columbia U.P. 1948
4. Nat. Bur. Stand. Computation Laboratory: Tables of Bessel functions of fractional order, Vol. 2. New York: Columbia U.P. 1948

Subject Index

MIX
Papier aus verantwortungsvollen Quellen
Paper from responsible sources
FSC® C105338
FSC
www.fsc.org

If you have any concerns about our products,
you can contact us on
ProductSafety@springernature.com

In case Publisher is established outside the EU,
the EU authorized representative is:
Springer Nature Customer Service Center GmbH
Europaplatz 3, 69115 Heidelberg, Germany

Printed by Libri Plureos GmbH
in Hamburg, Germany